Vital coalitions, vital regions

Vital coalitions, vital regions

Partnerships for sustainable regional development

edited by:
Ina Horlings

Wageningen Academic
Publishers

This publication has been made possible by:

Telos, the Brabant Centre for Sustainable Development in Tilburg, the Netherlands
telos@uvt.nl, www.telos.nl

TransForum, Zoetermeer, the Netherlands

TransForum encourages the necessary sustainable development of Dutch agriculture by linking it to its metropolitan environment.

ISBN 978-90-8686-141-5
e-ISBN: 978-90-8686-695-3
DOI: 10.3920/978-90-8686-695-3

Cover photo:
Artwork of Jan de Vries
in Lithoyen near the river Maas

Language-editing: Derek Middleton, Zevenaar, the Netherlands

First published, 2010

© Wageningen Academic Publishers
The Netherlands, 2010

This work is subject to copyright. All rights are reserved, whether the whole or part of the material is concerned. Nothing from this publication may be translated, reproduced, stored in a computerised system or published in any form or in any manner, including electronic, mechanical, reprographic or photographic, without prior written permission from the publisher:
Wageningen Academic Publishers
P.O. Box 220
6700 AE Wageningen
The Netherlands
www.WageningenAcademic.com

The individual contributions in this publication and any liabilities arising from them remain the responsibility of the authors.

The publisher is not responsible for possible damages, which could be a result of content derived from this publication.

Table of contents

Preface *Clarence N. Stone*	7
Chapter 1 – Introduction and research questions *Ina Horlings*	9
Chapter 2 – Creating capacity to act and sustainability in rural-urban regions: problem analysis *Ina Horlings*	17
2.1 Introduction	17
2.2 Regional development and planning in the Netherlands	18
2.3 Scenarios for sustainable rural development	21
2.4 Experiences with region-oriented policy and self-governance in the Netherlands	25
2.5 Introducing the concept of regimes	28
2.6 Problems with public-private co-operation in regimes	29
2.7 Can vital coalitions create capacity to act in regional development processes?	31
2.8 Creating sustainability in the interaction between regimes and vital coalitions	32
2.9 Summary and conclusions	34
References	35
Chapter 3 – New concepts for studying regional development *Julien van Ostaaijen*	41
3.1 Introduction	41
3.2 Urban regimes and vital coalitions: forms of policy networks	41
3.3 The urban regime concept	42
3.4 Vital coalitions	52
3.5 Regional regimes and vital coalitions	57
3.6 Summary	59
References	60
Chapter 4 – Pathways for sustainable regional development: strategies and vital coalitions *Ina Horlings*	63
4.1 Introduction	63
4.2 Methodology	64
4.3 Innovative ways of increasing the scale of land-based agriculture	65
4.4 Decoupling and industrial ecology	70
4.5 Multifunctional agriculture	73
4.6 Rural-urban integration	75
4.7 Types of coalitions	81

4.8 Contribution to sustainable development	82
4.9 Stimulating and hindering factors in regional development	83
4.10 Towards a new regional paradigm in rural-urban regions?	88
4.11 Conclusions	91
References	92

Chapter 5 – Vitality and values: the role of leaders of change in regional development — 95
Ina Horlings

5.1 Introduction	95
5.2 Methodology	96
5.3 The role of leadership in sustainable regional development	97
5.4 Leadership types and tasks	100
5.5. Leadership capabilities	102
5.6 Value-oriented leadership	103
5.7 Leaders of change in regional projects in the Netherlands	109
5.8 Empirical findings: a typology of leaders of change in Dutch regional projects	113
5.9 Conclusions: towards value-oriented leadership	119
References	122

Chapter 6 – New Markets in Heuvelland: coalition building and agenda setting — 125
Hetty van der Stoep and Noelle Aarts

6.1 Introduction	125
6.2 Issue development	128
6.3 Constructing coalitions without a sense of direction	142
6.4 The 'constructed power' of the provincial government	147
6.5 Self-organisation and synchronicity	151
6.6 Conclusion and discussion	154
References	156

Chapter 7 – Conditions for vital coalitions in regional development — 157
Julien van Ostaaijen, Ina Horlings and Hetty van der Stoep

7.1 Introduction	157
7.2 Vitality of the studied 'vital coalitions'	158
7.3 Conditions for vital coalitions	160
7.4 The interrelations between the conditions	166
7.5 Regime-coalition interplay and reflection on the theory	167
7.6 Recommendations	170
7.7 Summary	172
References	173

About the authors	175

Preface

Achieving co-operation around social purposes is one of the unending challenges of modern life. A business executive once remarked that, in a meeting if you want to get anything done, the first step is to have all of the lawyers and accountants leave the room. Co-operation that is based on neither the commands of a figure of authority nor on monetary compensation can be very powerful. It can also be hard to achieve. That is why authority and monetary compensation are so frequently employed. They are often readily available while voluntary co-operation may be uncertain. But formal authority is notorious for its weakness as a motivator and uncritical embracers of neoclassical economics notwithstanding, monetary payments are an imperfect source of motivation as well. Max Weber famously observed that individuals will die over a matter of honor but not a matter of money.

Research on the behaviour of members of the military in combat finds that the strongest bond among soldiers – the foundation of their willingness to die in battle – is not abstractions like patriotism but comradeship. Yet there are times when individuals do willingly die for a cause. Given the right conditions, the inspiration of a charismatic leader, or assurance that others too are committed to a struggle, people make extraordinary efforts and put their lives and livelihoods on the line.

There is much we know in a general way about social co-operation, but often we are uncertain about what applies in particular circumstances. This book takes the needed step of linking theory to concrete action. Co-operation across regional settings faces a number of similar challenges, but they are challenges that can take shape in various ways. Specifics matter, but they are usually instances of larger patterns we strive to understand. The concept of a vital coalition takes this challenge to the grassroots level and raises the question of how to get societal energy engaged by means of citizen interaction – interaction with the lawyers and accountants out of the room.

In past times of greater social, political and economic stability, immemorial custom could play a large part. For today, however, the creative destruction of capitalism and the elevation of innovation into a prime virtue have generally eliminated custom as a source of social co-operation. Furthermore, ours is a society of many impersonal relationships. Theologian Reinhold Niebuhr reminded us how moral codes and a sense of responsibility that hold in small-scale personal situation are nearly impossible to carry over into large-scale, impersonal situations.

The special challenge of the contemporary setting is that complexity heightens interdependence at the same time that it makes interdependence more difficult to decipher. Impersonality and strategic (amoral) calculations displace morality and social responsibility. The free-rider mentality does not flourish in a small, close community in the same way that it does in large settings, where relationships are more distant and a sense of oneness is harder to bring about.

Preface

Against this pessimistic view of the chances for widely practiced co-operation, we need to pose the optimistic view point that human beings are capable of social learning. They can observe past shortcomings and devise new arrangements to overcome the failings of the past. Efforts to create vital coalitions, learn from regime alternatives to business contracts and monetary purchases, and put into operation arrangements for large-scale social capacity also fall into the category of the humanly possible. It is promising practices in this vein that contributors to *Vital coalitions, vital regions* bring into consideration This is important work for regionalism and beyond.

Clarence N. Stone

Washington, DC

Chapter 1
Introduction and research questions

Ina Horlings

Many rural regions in Europe are undergoing a dynamic transition. Urbanisation, agricultural change, new patterns of production and consumption, and new societal demands are driving changes in the activities, functions and land use patterns found in these areas. The historical physical and sociocultural rural-urban divide is eroding. At the same time, the subjective experience of rurality is becoming more important to people for leisure, recreation and personal orientation. Once purely productive in function, rural areas are becoming increasingly consumer-oriented, providing new products and services for urban dwellers.

The complex processes affecting rural areas are proving difficult for the current institutions to manage, and the Dutch planning system is incapable of dealing with the many interrelated problems posed by the scarcity of space in the Dutch regions. At the governmental level, strong divisions remain between urban and rural authorities and between various 'vertical' sectoral policy domains operating in a 'business as usual' mode. A further steering problem is the 'administrative gap' at the regional level, which hampers an adequate development of rural-urban regions. Municipal policies, national sectoral policies and attempts by provincial authorities to fulfil an intermediary role all come together at the regional level. Although many governmental organisations are involved, there is no single body with exclusive decision-making powers.

Governments are trying to mobilise societal capacity in rural areas through forms of governance such as horizontal co-operation, co-production and negotiation between public and private actors. However, they are constrained by the limits of their managerial capacity and face institutional deadlock. Although such forms of governance create new multi-actor networks, they also lead to a diffusion of power and a lack of *capacity to act*. Niche innovations in the form of regional initiatives are held back by a glass ceiling of regulations and procedures, and encounter difficulties in setting new agendas for their region. To create more room for manoeuvre they co-operate with government authorities, but run the risk of being absorbed into current *regimes*.

There is a sense of urgency underlying the creation of new capacity to act. As the ability of governments to act in a hierarchical command and control mode diminishes, they seek a role as 'enabling power'. We will show that while the current arrangements used in Dutch rural policy strengthen government legitimacy, they are not effective enough. The new forms of governance do not adequately address the problem of the 'policy management gap'. In itself, the current shift from government to governance can be considered to be a positive development. Nevertheless, this process is hampered by clashing discourses, insufficient use of self-regulation and inflexible arrangements. These problems cannot be explained from the perspective of networks alone. To examine them properly we need to adopt a regime approach and a perspective on creating new enabling power.

This book addresses the pressing problem of how to create new capacity to act and get sustainable initiatives off the ground, initiatives that are now often smothered or constrained by the institutional context. We also deal with the problem regions face of creating spatial quality where different functions compete for space in the green environment. In many regions in the Netherlands and other parts of Europe the desire to combine economic development with landscape and natural quality leads to tensions and conflicts.

There is no single pathway to sustainable, regional development. Different scenarios and strategies can be distinguished, depending on the regional context and the actors involved. The opportunities for development depend on the extent to which regions can exploit regional assets to create a distinctive identity and can add value by forging new horizontal and vertical linkages. Expressions of this process of regionalisation are the production of new regional quality products, regional branding, new rural services and linkages between producers and consumers, new spatial designs and new alliances between agricultural sectors and other sectors.

In our opinion, these pathways require a re-orientation in the type of vertical and/or horizontal alliances or coalitions (both private-private, public-private and public-public) involved in agricultural production and rural development. Co-operation and interaction in regional processes in themselves are not sufficient; to attract public involvement they must have vitality. Modern citizens are not prepared to invest a great deal of time in public decision making if it is not made attractive to them. Decision-making processes must be embedded in a context that is inspiring and energising and offer realistic prospects of tangible results.

In this book we investigate whether *vital interaction* between actors and the creation of vital story lines as basis for new regional agendas can contribute to better regional co-operation and sustainable development. Vital interaction is a form of decision making in which co-operation between actors is energising and productive. By productive we mean that results are achieved; by energising we mean that the interactions give energy and inspiration to those who are involved. Vitality is a combination of productivity and energy. Vital interaction takes shape in an ideal situation in the form of *vital coalitions*. A vital coalition is *a form of active citizenship and self-organisation in which citizens and/or private or public actors take the initiative to act on behalf of a common concern or interest* (see Chapter 3). Empirical research done in a local urban context, based on the urban regime theory, shows that vital coalitions are informal forms of co-operation that can create new political power and capacity to act.

Vital coalitions can be seen as specific types of networks. There are three main differences from interactive policy making. The first is that not public, but private actors or actors in civil society have the initiative and take responsibility for public goals. Second, the focus is more on informal negotiation, dialogue and personal contact than on formal co-operation, with the aim of creating productive action and room for manoeuvre. Third, the goal is not so much the making of plans or policy, but the development of investment propositions, regional story

1. Introduction and research questions

lines or implementation strategies. Regional story lines can attract people to support a joint agenda, as Stone has shown in Atlanta (see Chapter 3).

A possible way to assemble vital coalitions is through innovative socioeconomic opportunities. The driving force for building coalitions is the combined investment possibilities for entrepreneurs who organise themselves around these opportunities within chains, clusters or regional multifunctional business communities. Examples of new coalitions that potentially contribute to more synergy between People, Planet and Profit can be found in several regions in the Netherlands and in other regions in Europe (see Chapter 4). Key actors in the process of building coalitions in the cases are *leaders of change*, initiators that have the motivation, energy and passion to align people around new agendas (see Chapter 5). The projects focusing on the integration of rural-urban development in particular illustrate what is called by the OECD a 'New Rural Paradigm'. The core of this new paradigm is the need for a closer linkage between the rural and urban economy and to integrate rural development more closely with regional development in general. This includes a new, multisector, place-based approach to rural development which seeks to identify and exploit the varied development potential of rural regions through new industries, such as rural tourism, manufacturing and ICT. According to the OECD, this 'new rural development' expresses a shift from subsidy-driven development to development through investments in different countries. The key elements are greater competitiveness of rural areas, based on the valorisation of local assets and the exploitation of unused resources, and the influence of various sectors. The new paradigm requires the development of new collective governance arrangements to better integrate a broad range of state and non-state actors, horizontally at both the central and local levels, and vertically across all tiers of government.

But the question that remains is how to establish these arrangements. The central research questions explored in this book are:

> *How vital are public-private coalitions that aim to contribute to sustainable innovations in rural-urban regions in the Netherlands?*
>
> *How do current regime conditions hinder or facilitate the mobilisation of vital coalitions?*
>
> *What new conditions are needed to organise vital coalitions?*

The goal is to analyse formal and informal networks as a source of regional innovation and the stimulating/hindering role of current and possible future institutional (or regime) conditions. The goal is to develop conceptual and organisational tools to help vital coalitions to mobilise a new capacity to act in rural-urban regions. Eight varied regional cases are analysed, illustrating different pathways to sustainable development (see Figure 1.1):

- The Overdiepse Polder in the province of Noord-Brabant.
- Sjalon, a farmers initiative in the province of Flevoland.
- The New Mixed Business, a cluster of agricultural businesses which will be located near Venlo in the province of Limburg.

- The Northern Frisian Woods, a combination of four farmers' associations in the province of Friesland in the north of the Netherlands.
- The Regional Innovation Centre in the Arkemheen Eemland National Landscape near the city of Amersfoort.
- Waterland, a part of the Laag Holland National Landscape in the province Noord-Holland in the west of the Netherlands.
- Het Groene Woud, a National Landscape in the province of Noord-Brabant.
- Heuvelland, a National Landscape in the southeastern tip of the Netherlands.

1. Noordelijke Friese Wouden (national landscape)
2. Sjalon
3. Waterland (part of national landscape)
4. Eemlandfarm (national landscape)
5. De Overdiepse Polder
6. Het Groene Woud (national landscape)
7. New Mixed Business
8. Heuvelland (national landscape)

Cartography by Jan Edwards

Figure 1.1. The regional cases illustrating different pathways to sustainable development.

The process of coalition building and agenda setting for one case study, Heuvelland in Limburg, is described in detail and analysed in depth (Chapter 6).

In this book we analyse empirical findings using a theoretical framework that has been successful in the US, but has not yet been applied to regional development in Europe. The conceptual framework provided by urban regime theory offers insights into the complexities of regional practice with its competing interests, and the existing planning and institutional context. The concept, which is explained in Chapter 3, allows us to study the dynamics of innovations at the regional level within the context of various governance structures and relevant processes, such as the organisation and functioning of the public policy system and spatial developments. Based on the outcomes of studies using urban regime theory in the urban context, we postulate that mutual agenda setting, mutual resources, a mutual action space and effective modes of alignment between (public and private) actors are also the building blocks of innovative vital coalitions in rural-urban regions.

By analysing coalition building from a combined theoretical-practical perspective, and by comparing innovative regional projects, we can obtain interdisciplinary insights into the conditions that favour the formation of vital coalitions in different institutional contexts. The research offers insights into the factors influencing the success or failure to build coalitions and transform regimes. The ambition is to give a perspective on transition processes by identifying the conditions under which informal networks might function as a breeding ground for new resilient regional regimes that facilitate sustainable regional development.

The literature on regional development is extensive. However, many publications are either purely theoretical or limited to empirical case studies, including concrete lists of what or what not to do. The aim of this book is to bridge the gap between theory and practice and develop greater understanding of the possibilities for enabling the transition towards sustainable regional development. Our ambition was to compile a book that tells a story and analyses different aspects of vital coalitions: the theoretical background, the strategies pursued by actual initiatives when dealing with current regimes, the role of leaders of change as key actors, the process of agenda setting and coalition building, and the conditions for vital coalitions.

The central question explored in this book is divided into several subquestions, which are examined in the following chapters. An outline of the book and the research questions is given below.

In Chapter 2 the following questions are explored: *what problems are encountered in Dutch rural-urban regions and how can capacity to act be created to realise more sustainable development?* The starting point in this chapter is an evaluation of regional development in the Netherlands over the last few decades and the problems that occur, especially the obstacles to public-private co-operation. The search for innovative approaches in rural-urban regions has tended to be locked within traditional arrangements. New regional actors, such as non-agricultural entrepreneurs, and urban representatives, face difficulties when trying to introduce their own problem definitions, participate in rural decision-making processes and co-operate with institutionalised governments. The chapter introduces the main concepts used in the book, such as sustainability, regimes and vital coalitions. The goal is to show how the episodic power of self-organised initiatives challenges structural forces and how they are supported or hindered by regimes.

The central questions in Chapter 3 are: *how can the concepts of urban regimes and vital coalitions be used to study public-private initiatives in a regional setting, how do these two concepts relate to each other, and what are conditions for vital coalitions?* This chapter builds on the basic themes of the urban regime theory, drawing on examples of vital coalitions and experiences with the urban regime theory so far. The question is whether the important preconditions for successful vital coalitions identified in the case studies of urban development in the US and recent local urban development in the Netherlands also apply to forming vital coalitions in rural-urban regions.

The central questions in Chapter 4 are: *which types of coalitions can be identified in the context of Dutch regional development, to what extent do they introduce new sustainable issues and agendas on the regional scale, and which obstacles do they face?* The Dutch context of regional development is explained and four scenarios for regional development are described: (1) an increase in the scale of agricultural production in both intensive and extensive systems; (2) an ongoing decoupling of agricultural production from the physical environment; (3) multifunctional agriculture; (4) rural-urban integration. These are illustrated by eight case studies, in which we describe which forms of coalitions can be distinguished, how these coalitions reflect different scenarios for sustainable development, the agendas they introduce, and what obstacles they face in regional co-operation. We also address the governance implications of the diversification of the countryside.

The central question of Chapter 5 is: *what is the role of leaders of change in regional development?* The 'human factor' – the characteristics of individuals – is an important condition for energising and productive coalitions. The key actors in the eight regional cases it are private initiators, who play a leading role. We build a framework to analyse these initiators based on an inner and an outer dimension, and an individual and collective dimension. The description of these dimensions shows how leaders of change function in vital coalitions, their characteristics, their passion and motivation and their roles in regional co-operation. We also examine the strategies they follow in trying to create capacity to act in interaction with the current institutional context.

The central question of Chapter 6 is: *how did the New Markets initiatives develop agenda-setting capacity for sustainable regional development in Heuvelland?* Vital coalitions have been described as coalitions with a capacity to achieve outcomes, a capacity to act. In this chapter 'capacity to act' is operationalised as agenda-setting capacity, the ability of private-public initiatives to construct an appealing agenda that mobilises support. The New Markets in *Heuvelland* project provided information for an in-depth case study of the agenda-setting capacities of various initiatives. The analysis focused on issue framing and coalition-building strategies in the agenda-setting process. From our findings we make some recommendations for the organisation of vital coalitions. The chapter argues that vital coalitions should be viewed as a process of coalition building in which the creation of a powerful and appealing story line is more important than keeping participants in the coalition.

1. Introduction and research questions

In the final chapter (Chapter 7) we reflect on the significance of the empirical results and draw conclusions based on the previous chapters. We also evaluate the urban regime theory as a framework for understanding regional development and try to answer the central questions. We compare our research results with results from earlier research into urban development processes to answer the question: *do the mechanisms, used successfully to build coalitions and create the capacity to act in those cases, also apply in the rural and urban context?*

This research was carried out as a two-year scientific project, partly financed by TransForum. TransForum encourages the sustainable development of Dutch agriculture by linking it to its metropolitan environment, using a combination of knowledge, networking and resources. It helps to create coalitions of knowledge institutes, entrepreneurs, governments and civil society organisations, which together develop innovative forms of economic activity that are profitable, respect the environment and improve human and animal welfare (www.transforum.nl).

The project was carried out by Telos, in co-operation with Wageningen University and Research Centre (WUR) and the University of Tilburg (UvT). The interdisciplinary project team consisted of three researchers:
- Ina Horlings (project leader and senior researcher in rural and regional development at Telos);
- Hetty van der Stoep (researcher, Land Use Planning Group, Wageningen University and Research Centre);
- Julien van Ostaaijen (researcher, Tilburg School of Politics and Public Administration, University of Tilburg).

Professor Noelle Aarts contributed to the project during the last phase as co-author of Chapter 6.

We thank the members of the steering group, who gave us useful advice during the project: Professor Hans Mommaas (Telos), Professor Adri van den Brink (WUR), Dr. Jan van Tatenhove (WUR) and Professor Frank Hendriks (UvT).

We hope that this book offers planners, policy makers, researchers, project managers and students insights into how effective regional co-operation and new sustainable regional agendas can be established, and how they can respond to the current challenges of urbanisation, recreation, agricultural development and climate change, thus contributing to sustainable development in rural-urban regions.

Chapter 2
Creating capacity to act and sustainability in rural-urban regions: problem analysis

Ina Horlings

2.1 Introduction

As we stated in Chapter 1, a transition is taking place in rural-urban regions. New functions are competing for the scarce rural space, new actors are entering the rural arena and new challenges, such as climate change and urban demands, have to be faced. Achieving a more sustainable development of these regions will require a *capacity to act*. The possibilities for government authorities to act in a hierarchical command and control mode are diminishing; instead, they seek a role as *enabling powers*, but how can they fulfil this role? Innovation occurs through regional initiatives, but these initiatives come up against a glass ceiling of regulations and procedures, which make it difficult to set new agendas for the region. To create more room for manoeuvre, such initiatives co-operate with government, but they then face the risk of being absorbed into current *regimes*.

In this book we investigate whether *vital interaction* between actors and the creation of *vital storylines* can contribute to better regional co-operation and sustainable development. We set out to discover whether *vital coalitions*, new forms of informal partnerships based on vitality, can be found in rural-urban regions and how these coalitions are hampered or stimulated by current regimes.

As described in Chapter 1, the core questions of this book are divided into several subquestions, which are dealt with in the different chapters. This chapter deals with the following subquestion:

> *What problems are encountered in Dutch rural-urban regions and how can capacity to act be created to realise more sustainable development?*

First, we describe the regional development and planning context in the Netherlands (Section 2.2) and the concept of sustainable development and scenarios for sustainable development in rural areas under urban influences (Section 2.3). Dutch regional policies contain different arrangements for facilitating the shift to sustainable development, based on the change from government to governance. More specifically, in recent decades in the Netherlands new forms of region-oriented policy and self-governance have emerged. These new activities, methods and prospects bring their share of problems as well, as we will see (Section 2.4). The concept of regimes is helpful in analysing these problems. This is an analytical approach designed to throw light on the obstacles to regional co-operation. We introduce the urban regime theory, developed in the context of urban renewal in the US, which offers a fresh perspective on regional complexity (Section 2.5). Based on this theoretical framework, we explain how public-private co-operation is hampered in rural regimes (Section 2.6). We also examine whether a solution to these obstacles may be found in specific networks called vital coalitions, a form of vital interaction between regional actors characterised by energy and productivity. Our hypothesis is that vital coalitions can create a capacity to act in regions that have become gridlocked by current procedures and regulations (Section 2.7). To throw more light on this hypothesis we need a greater understanding of the relations between vital coalitions and regimes (Section 2.8). Finally, in Section 2.9 we draw some conclusions.

2.2 Regional development and planning in the Netherlands

The Dutch rural area is man-made. The polders in particular reflect the rational place-making philosophy of spatial planning. The quality of the Dutch landscapes has been influenced not only by land reclamation, but also by rural land development and restructuring programmes. Agricultural land was reallocated and consolidated after the second World War, creating the conditions for mechanised, intensive and highly-productive agriculture. Since then technological innovations have led to the decoupling of agricultural production from the local environmental conditions that strongly influenced farm production (Van der Ploeg, 1992). Examples are greenhouse horticulture, which is decoupled from the soil, the 'closed' animal housing systems in intensive pig and chicken farming, and genetically modified crop varieties that are resistant to herbicides. We also see a global and social decoupling. Raw materials for food production, like sugar from beets and cane or of chemical origin, can be exchanged globally by the food processing industry as a result of technological developments (Horlings, 1996: 26). The family structure of farms is also under pressure from the expanding scale of production. The negative sides of the dominant post-war development model are well known: falling incomes, environmental damage, problems with animal welfare, loss of landscape quality and biodiversity, and loss of confidence in food quality (Marsden, 2003; Wiskerke and Roep, 2007).

For decades it remained possible to manage the countryside, as long as spatial planning and agriculture had relatively parallel interests and agendas. Societal demands on the countryside did not conflict with the actual development of regions. The 'Green Front', a national agricultural regime in which policy, research and extension services worked in unison, had a strong influence on the agendas in rural areas, and for decades agriculture and countryside development were almost synonymous. The agrarian sector successfully shaped rural policy and rural space (Frouws, 1989).

This changed when the issues of sustainability, urbanisation and increasing societal demands all came together. From the 1980s rural residents were given increasing influence in rural land development projects (Driessen, 1990). Environmental problems like acidification, eutrophication and manure surpluses became manifest, while the growing scale of various rural functions (agriculture, water, housing, infrastructure) threatened the quality of valuable small-scale landscapes. Specific landscapes, characteristic land uses and their particular flora and fauna disappeared (RIVM, 2002). These developments led to friction between urbanisation and spatial quality (Ritsema van Eck and Farjon, 2008). The environmental problems associated with agriculture, as well as animal diseases and food scandals, affected the image of farming and the interests of market-oriented agriculture clashed with the public perception of the countryside and animal welfare. During the outbreak of foot and mouth disease, for example, citizens and farmers protested against the killing of non-vaccinated cows in the Netherlands (Van der Ziel, 2003).

On the spatial level, the urban-rural dichotomy eroded in the *metropolitan landscapes* (Wiskerke, 2007), where urban and rural activities are becoming increasingly intermingled. These areas are undergoing a transformation. Once dominated by agriculture, they are now becoming increasingly dominated by the leisure demands of the growing urban middle class, often described in symbolic and cultural terms (nature, authenticity, freedom, etc.). To meet their recreational demands, these urban landscape consumers want space, peace and quiet, and attractive landscapes. The notion of a *consumption countryside* illustrates this transformation (VROM-raad, 2004: 227). Rural-urban areas have become network societies, where local and international production and consumption are connected in a complex system, whereas governance implementation is still organised along sectoral lines.

Urban development exerts a growing influence on these regions, as the housing market becomes more regionalised and the spatial patterns of citizens' recreational behaviour cross urban boundaries (Mommaas *et al.,* 2000). Instead of lying outside the city, rural areas now find themselves lying 'in between' urbanised areas. Urban-rural boundaries are disappearing as the expansion of urban activities and the housing market bring about a suburbanisation of work and supply functions (Verwijnen and Lehtovuori, 1999).

This blurring of rural-urban boundaries and the rise of new, more urban demands is changing the countryside in a large part of the Netherlands as new functions, new inhabitants and new practices take root in rural areas. These changes are not only reducing the power of rural actors, but also diversifying and complicating the rural arena, raising questions about conceptions of rural space (Frouws, 1998; Marsden, 1998). Agriculture has to adapt to this new situation and, following the outbreak of foot and mouth disease, the swine fever crisis and food scandals, also has to re-earn its 'licence to produce'. New industrialised forms of agriculture clash with people's values, as illustrated in 2007 and 2008 by protests in the Netherlands against 'mega animal sheds'. Consumers are putting greater importance on the origin, the experience and the 'story' of agricultural products, and their interest in sustainable production methods and animal welfare is growing.

The problem in Dutch agriculture is that its autonomous development is out of line with the spatial, landscape and governance context. Moreover, people have a range of different ideas about what spatial quality is, which leads to conflicts of interest. Agriculture can no longer play its role as the undisputed guardian of rural areas, and new organisational principles are needed in order to diversify and revitalise the rural economy, while strengthening the highly valued ecological and sociocultural qualities of rural areas (Mommaas and Janssen, 2008: 25). The question is, which new agricultural scenarios can create a fit with societal demands and again deliver agricultural practices that maintain landscape quality.

Environmental, social and regulatory problems are closely interrelated, and are bringing about a process of rural transition in these regions. The increase in the scale of various rural functions (agriculture, water, housing, infrastructure) is eroding regional boundaries. For this reason, spatial planning faces difficulties in fulfilling its integrating function. A complicating factor is that what constitutes an adequate area for regional development can be quite variable, depending on the geographical environment, natural resources and amenities, skills and infrastructure (OECD, 2006: 114).

Traditional regional planning procedures can only provide part of the solution to the integration challenge and the complexity of rural issues is proving difficult for the current institutions to manage. One of the problems is that the 'administrative gap' at regional level hampers an adequate development of rural-urban regions. Municipal policies, sectoral national policies and attempts by provincial councils to fulfil an intermediary role all come together at the regional level. Although many governmental organisations are involved, there is no single body with exclusive decision-making powers. This is perceived as constituting a 'gap' in regional governance (Van den Brink et al., 2006). Hajer and Zonneveld (2000) argue that the Dutch planning system is not fit to deal with the complex and interwoven problems posed by the scarcity of space in the Dutch regions. Van den Brink describes the example of self-referentiality of government, the 'urban-rural divide', in the context of the rural-urban landscape. This urban-rural divide is caused by the division of responsibility for spatial policy between urban planning departments, housing departments and rural development departments, not only at the national level but also at provincial and municipal levels (Van den Brink et al., 2006).

The planning problems become even more complex within the wider context of the reorientation of governmental roles. As a reaction to failing governmental legitimacy and effectiveness in the past, government authorities have adopted new approaches based on the assumption that they no longer have the means necessary to manage and control the physical, economic and social environments on their own and are becoming increasingly dependent on the knowledge and involvement of groups within society. Technical expertise is insufficient; social expertise is indispensable (Stoker, 2003). This is what some have come to call the *horizontalisation* of public administration, which has been the topic of much discussion over the last ten to fifteen years. It is a very broad approach and encompasses a number of different lines of thought (Torfing et al., 2003), all based on the common proposition that the classic hierarchical model of public administration does not work well enough, and that a number of forms of 'horizontal arrangements' have arisen in its place. Even though government authorities have always relied on other agencies to aid them in realising national objectives, the central role of 'the state' is decreasing (Jessop, 1997).

Concepts like network management, interactive decision making and co-production all relate to this horizontalisation. They are manifestations of more fundamental changes in the relationship between government and citizens (Lovan, 2004). 'Governance' as an alternative to 'governing' consists of formal and informal regimes based on interaction, partnership and co-operation between public and private actors, or the self-regulation of the latter (Rhodes, 1997). It has to do with a diminution of the role of the state, with market forces taking over some functions and producing some goods and services previously seen as state prerogatives, and with a shared commitment to resource allocation and conflict resolution (Schmitter, 2002). The rise of governance is seen at the blurring of boundaries between and within the public and private sectors. According to Stoker (1998), this is believed to restructure collective action and to re-establish an order based on social co-operation, mutual interest and accommodation. The changing power relations between state, market and civil society lead not only to new styles of governing, but also solidify into new relations and arrangements. Examples of these new relations between the state and civil society can be found in the context of rural regional policy.

Besides horizontalisation, the trend of decentralisation, in which new responsibilities are delegated to subnational tiers of government, is also relevant to regional steering. At the same time, greater attention is also given to place-based policies, which means there is an increased focus on the role of local entities in the implementation of such policies. Bottom-up approaches are being encouraged in several countries. However, in various European countries, local and regional partnerships are facing a number of potential obstacles, such as the complexity, rigidity and fragmentation of national and supranational policies, which affect rural development (OECD, 2006: 114, 127).

Recent developments in European rural-urban regions relate to what the OECD (Organisation for Economic Cooperation and Development) calls 'a new rural paradigm' (OECD, 2006), a shift from subsidy-driven development to development through investments. According to the OECD, this requires new governance arrangements in which public agents do not dominate, but where there is strong representation by the private business sector to capitalise on new opportunities and resources generated by public or collective action (OECD, 2006: 114).

The projects described in the following chapters can be seen as concrete 'bottom-up' answers to the trends and tensions described above. The projects are forms of self-organisation that illustrate new agendas for rural-urban landscapes. However, the expression of these agendas is strongly influenced by the regional context.

A complicating factor is that regional actors are part of regimes – interwoven combinations of dominant ideas, actors, agendas and rules about development. New regional agendas can be a source of friction with the current regimes. Our central hypothesis in this book is that future scenarios for rural-urban regions need a common basis for co-operation in the form of vital coalitions between actors. We will investigate whether vital coalitions can be observed in practice, whether they contribute to sustainable regional development, and what the conditions are for vital interaction at the regional level. Sustainable development is defined in this book as a balance between economic, sociocultural and ecological issues without causing negative effects in space (other countries) or time (future generations) (Telos, 2002). This concept is explained in the next section.

2.3 Scenarios for sustainable rural development

Globalisation may make people feel more uncertain about the future. Bauman (2005) refers to this as 'the liquid life', a life in which everything is possible and people have to find their own ways to organise their lives. People experience the uncertainty of continuous change, suffer from that and are afraid of being left behind. Life is like 'skating on thin ice'. In these dynamic times people need an anchor point to which they can turn. Sustainability can offer that reference and a sense of direction (Horlings *et al.*, 2009a: 26).

Since the presentation of the report *Our common future* (WCED, 1987) it has become clear to many that we have to be more aware of the consequences our actions have on the earth, people elsewhere and future generations. Sustainable development is a development process that seeks to find a balance between the sociocultural, ecological and economic stock, or, in popular terms, between People, Planet and Profit (Figure 2.1). Strengthening one stock must not be at the cost of another stock, not now, not later and not in other regions or parts of the world. It is a broad and integrated approach with both a strategic and a normative dimension. The strategic dimension refers to the responsibility to make short-term decisions from the long-term perspective of sustainability. The normative dimension takes the effects on future generations and other geographical levels into account (Telos, 2002). Critical change and innovation processes do not occur at the points, but in the heart of the triangle depicted in Figure 2.1, where interaction between People, Planet and Profit has real consequences. This is the fourth dimension of sustainability. It is more important than the other domains and always depends on the regional context. In times of economic stagnation, social and environmental goals are often seen as obstacles. Sustainability is then associated with constraints on new developments, but investments in sustainability can lead to new innovations and create new opportunities.

Sustainable development has recently become a central theme in regional development (Pike *et al.*, 2006). Today's post-industrial society challenges regions to stimulate economic development by investing in ecological and cultural qualities, to develop eco-economic innovations based on regional qualities and the availability of resources such as water, landscape quality and forests. Regional qualities and resources can lead to changes in production and new investments (Mommaas and Boelens, 2006: 12). Strengthening the uniqueness of regions makes them less vulnerable to the pitfalls of globalisation, such as overexploitation and current competition between regions.

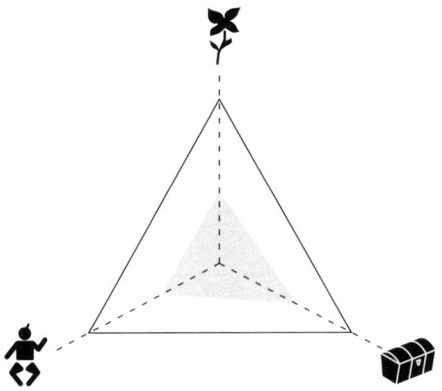

Figure 2.1. The sustainability triangle.

2. Creating capacity to act and sustainability in rural-urban regions

There is no single path to sustainability. The projects described in the following chapters illustrate a variety of perspectives on agricultural and rural development in the Netherlands, which are based on different discourses. There are many definitions of 'discourse', but here we define discourses as 'an organised set of social representations, the terms through which people understand, explain and articulate the complex social and physical environment in which they are immersed' (Frouws, 1998). By giving meaning to the world, human agencies construct discourses, but at the same time 'existing' discourses mediate this meaning-given process. Hajer, whose definition of discourse is often quoted, emphasises the relation between discourse and practice: 'A discourse is an ensemble of ideas, concepts and categories through which meaning is given to social and physical phenomena, and which is produced through an identifiable set of practices' (Hajer, 1995, 2006). Discourses are the heart of a regime, structuring regulation, resources and the formation of actor coalitions (Arts and Leroy, 2006). Actors can adapt to these discourses or can introduce new ideas, creating tension with the current regime.

Different discourses have been identified in the field of agriculture and rural development. Frouws identified the agri-ruralist, utilitarianist and hedonist discourses (Frouws, 1998). He provided an interesting analysis of perspectives on the countryside, but never linked this to issues of sustainable development (Hermans *et al.*, 2010). Marsden (2003) identified three separate agrarian models for rural development and links them specifically to sustainable development issues. The first model is the agro-industrial model, which reflects the faith in free competition and is based on partial deregulation of markets. The second model is the post-productivist model, based on the idea that rural space and nature become consumption goods for the urban population. In Marsden's view, only his third model claims to be sustainable, the alternative sustainable rurality model, which he derived for definitions of agro-ecology. At the core of the concept of sustainable rural development is multifunctional agriculture. As Van der Ploeg *et al.* (2002: 12-13) explained, agricultural activities are deepened, transformed and expanded by linkages and associations with new actors and agencies. Sustainable rural development suggests the potential symbiotic interconnectedness between farms and the locale.

The challenge is to realise new sustainable scenarios for the future of agriculture. Since rural-urban boundaries have eroded, TransForum, a Dutch organisation which stimulates sustainable agriculture with new knowledge, has introduced the concept of *metropolitan agriculture* as a new pluralistic concept and a promising scenario for the future. Metropolitan agriculture is a system of agro-production designed for densely populated delta regions like the Netherlands; a system that responds to the changing and competing demands of the metropolitan society, based on new, intelligent connections between producers, sectors, resources, flows (energy and waste), triple-P-values and stakeholders. It makes use of and creates synergy between the unique selling points of the metropolitan environment (logistics, networks, consumer trends, flow of knowledge, etc.).

In our view, four trends or scenarios for rural development can be observed. Although these have been identified in the Netherlands, we can find these scenarios in many European agricultural landscapes under urban influences. These scenarios include elements of Marsden's model, such as post-productivism, market orientation and agro-ecology. In our opinion, however, none of the scenarios can be labelled as being exclusively sustainable, because this depends on the actual production methods used in practice, as we will see in the cases in the following chapters. These four scenarios are described below.

1. *Increase in scale of agricultural production*
   The focus in this scenario is on the enlargement of soil-based agricultural enterprises, such as horticulture and the dairy sector. Key elements are mass export- and market-oriented production and the use of external inputs. Increase in scale can lead to 'social decoupling', an erosion of the family-based structure of agriculture. This scenario does not reflect a sustainability strategy as such; however, by increasing the farm scale, the production costs as well as the use of artificial manure per kilogramme of product can be lowered. In this sense this strategy can be seen as an attempt to combine Profit (continuity and efficiency of farming) and Planet (less negative environmental effects) goals.

2. *Development towards an ongoing decoupling of agricultural production and the environment*
   This scenario can be seen as a continuation of the paradigm of rational, mechanised and intensive agricultural production. Over the last century the relationship between farms and their surroundings has disintegrated, which is termed 'decoupling'. In future, an ongoing process of decoupling is expected in the form intensive farming on industrial sites and new types of agro-industrial production. This can be adapted for sustainability, using industrial ecology principles, by clustering (multisector) intensive farm businesses at one location. The advantages of this include lower transport costs, more efficient use of energy and the exchange of resources and waste streams between sectors. However, it leads to growing friction with the People dimension of sustainability: people demand the maintenance of valued cultural landscapes, while farmers wants space to develop their 'megafarms'.

3. *Multifunctional agriculture*
   Multifunctional agriculture is a form of farming that involves new agricultural products as well as functions and services, such as nature and landscape management, the production of regional quality products, organic production, tourism, healthcare and water management. The goal is to develop new sources of income as a response to the price/cost squeeze on agriculture, but also to maintain the quality of agricultural landscapes by preserving such values as tranquillity and open space. Sustainable goals focus on landscape and nature protection, water management and a lower input of chemicals and minerals. European rural policy, implemented via instruments such as 'regional development plans' and the INTERREG and LEADER+ programmes, focuses mainly on this perspective.

4. *Rural-urban integration*
   This 'post-productive' perspective can be seen as a network approach that defines rural landscapes as consumption spaces, paying attention to societal demands, establishing multisector networks and designing new concepts based on rural-urban alignment and alliances between agricultural and non-agricultural entrepreneurs. In this scenario, new actors enter the rural policy arena (VROM-raad, 2004). The goal is to attract new investors, who function as drivers for landscape quality.

The first two scenarios are in line with the instrumental national regime that was dominant in the Netherlands for decades, although new environmental goals have been introduced since the 1980s. After the Second World War, research, extension and policy were strongly connected and generated a powerful agenda, planning instruments and research programme. European agricultural policy encouraged the scaling up of production.

This regime ceased to function properly when environmental came to the fore and gave way to a struggle between discourses. Under the umbrella of sustainable development, the struggle over the future of the Dutch countryside has intensified. The concept of sustainable agriculture has not led to a unified overarching vision for the future (Hermans *et al.*, 2010), although discourse coalitions have been shaped successfully in some regions for more than an one election period (see for example Mommaas and Janssen, 2008).

In the following sections we will see how government authorities have tried to manage rural development at the regional level in recent decades. We argue that there is an urgent need for new forms of regional arrangements and coalitions.

2.4 Experiences with region-oriented policy and self-governance in the Netherlands

Various types of formal and informal reflexive policy making in rural regions can be seen as an expression of new relationships between the state, the market and civil society. Since the 1990s these new relations have led to new kinds of policy, such as region-oriented policy and self-governance. Region-oriented policy (*gebiedsgericht beleid*) functions as a kind of network governance, a form of interactive development strategy in which government initiative is the starting point. Self-governance can be seen as bottom-up development in which regional initiatives take the lead.

The experiences with interactive policy making and region-oriented policy making are not all positive. For example, Pröpper and Steenbeek (2001) refer to the risks of passing on responsibilities to other organisations, lack of efficiency and interorganisational conflicts. Intensive negotiation and consensus-seeking processes with a wide variety of interest organisations take up much time and can lead to endless conflicts of interest. Interactive policy making that aims to reach a consensus may actually stifle innovative ideas (see also Van Stokkom, 2006). Participants can fall into old patterns of behaviour and pursue their own agenda or strategic interest, which can put agreements under pressure.

In the 1990s several forms of region-oriented interactive policy were developed:
- 'ROM areas' (*ROM-gebieden*) for the integration of spatial and environmental policies;
- 'Valuable Cultural Landscapes' (*Waardevolle Cultuurlandschappen*);
- 'Rural Restructuring Areas' (*Reconstructiegebieden*) in the eastern and southern parts of the Netherlands.

The ROM policy (in Dutch, ROM stands for *Ruimtelijke Ordening en Milieu*, spatial planning and environment) was designed within the framework of the fourth national policy on spatial planning (Ministerie van VROM, 1989) to integrate planning and environmental policies at the regional level. From the experiences in ROM areas such as *Schiphol* and *Gelderse Vallei* (Gijsberts, 1995) the conclusion can be drawn that the ROM policy, which set out to integrate policy measures by negotiating with different stakeholders, turned out to be an instrumental-technical approach that was institutionalised in traditional arrangements. The result was that environmental measures were weakened during the process (see also Ministerie van VROM, 1998).

Under the Valuable Cultural Landscapes policy, national government and regional authorities sought to maintain and develop the qualities of these valuable agricultural regions (nature, landscape, recreation) as well as pursuing economic goals. The Valuable Cultural Landscapes policy focused on regional processes and the stimulation of bottom-up initiatives, which led to new alliances between actors. However, the informal relations and new ideas conflicted with regional institutional processes, such as spatial planning procedures (Novioconsult, 1998; Pleijte *et al.*, 2000).

New regional forms of interactive policy over the last few years can be seen in the Rural Restructuring Areas in the Netherlands. The restructuring programme was launched mainly in response to the environmental problems caused by intensive livestock farming. During the process the aims broadened into an integrated approach to all rural problems, economic, social and ecological. Restructuring committees were formed that included representatives of the many actors involved in the process: regional and local governments, farmers organisations and a broad range of societal organisations. Broad co-operation between these private and public actors led to consensus building and the production of restructuring plans. An evaluation of the restructuring plan for 'De Peel', however, shows that opportunities for the self-organisation of concrete initiatives were not adequately exploited (Haarmann, 2004). Co-operative dealings occurred mostly within the restructuring committees in the form of traditional arrangements and negotiating processes with the well-known interest groups. This meant that the process of consensus building was not open to the input of innovative solutions from new actors, who were not able to participate.

In a review of these area-based initiatives Frouws and Leroy (2003) state that region-oriented policy is an exercise in social mobilisation, consensus building and resolving conflicts. New local and regional coalitions were formed, but most were temporary, and in their view the policy does not constitute a thorough renewal of governmental style. In her dissertation, Boonstra (2004) concludes that region-oriented policy has triggered intensive co-operation between governmental institutions and NGOs, mobilisation of resources, joint problem definition and action, and insight into constraining legislation. But in spite of these accomplishments, the potential of this policy with regard to public support, participation, mobilisation and integrating capacity has only been partially realised. Opportunities to mobilise new actors, resources and discourses are being missed. In addition, representatives of the NGOs involved tend to remain stuck in the same old patterns (Boonstra, 2004).

Based on the literature on different forms of interactive rural region-oriented policy, we draw the conclusion that governments have tried to mobilise social capacity in rural areas through horizontal co-operation, forms of co-production and stimulating negotiation between public and private actors. However, they run up against the boundaries of their management capacity and the implementation of such policies is beset by problems caused by clashing discourses, dominant power relations, a selective choice of actors and obstacles to co-operation between interest groups. *The problem is that actors try to solve new rural problems with new styles of governance, but using traditional coalitions* (see also Horlings, 2009b). In particular, this leads to problems with regional co-operation. Regional development requires the involvement of innovative ideas and partners, but public and private actors face the risk of being incorporated into 'solidified arrangements' not suited for tackling the new challenge.

The governance problems are heightened and made more complicated by what can be described as the rural-urban divide. The rural-urban divide is caused by a division of spatial policy between departments for urban planning and housing and for rural development, not only at the national level but also at the provincial and municipal tiers of government. Van den Brink *et al.* (2006) point out that there is also a lack of commitment from big municipal authorities when it comes to managing open spaces. Meanwhile, an increasing number of new players have emerged in the rural-urban landscape, as agricultural stakeholders give way to nature conservation organisations, water boards, recreation entrepreneurs, etc. These new players, which often have more 'urban' backgrounds, are becoming more influential because of the financial resources at their disposal. The factors mentioned above created a situation where no actor can monopolise the planning and management of rural-urban areas because power is diffuse.

The forms of region-oriented policy described above are illustrations of the transition from governing to governance. According to Pellizoni (2004), this transition is also visible in the development of voluntary regulation. Rhodes (1997) also sees the self-regulation of private actors as an element of governance. It creates opportunities for 'bottom-up' groups within civil society to take responsibility for sustainable development in their own region and to participate in regional co-operation.

The key characteristics of the types of self-governance in the Netherlands are:
- The development of voluntary instruments such as covenants, contracts and regional agreements between public and private actors.
- A focus on flexible policy solutions, experiments and 'changing the rules of the game'.

An example of self-governance in the Netherlands is the rise of agri-environmental associations consisting of groups of farmers and citizens working together on environmental, nature and landscape management (Melman, 2003; Oerlemans *et al.*, 2004). The most successful regional self-organising groups in the Netherlands have the knowledge and professionalism to co-operate with the government (Horlings, 1996).

Many innovative associations try to establish solid and lasting relations with government authorities by entering into contracts, or covenants. However, in the current situation several institutional barriers obstruct the development of regional co-operation. These barriers include a narrow focus on sector-based rather than integrated rural policy, European regulations, existing power relations and a bureaucratic attitude by civil servants. The institutional context can incorporate private actors into a decision-making process and the resulting tension creates a major dilemma for private actors: to obtain more room for manoeuvre they have to participate in institutional structures, but this can involve high 'transaction costs', with the attendant risk of a loss of vitality (see Horlings, 2009b).

2.5 Introducing the concept of regimes

Experiences with the two types of policy making described above – region-oriented policy and self-governance – show that although the current arrangements and partnerships used in rural policy strengthen governmental legitimacy, they are not effective enough. At the regional level, the 'policy management gap' is not effectively addressed by these forms of governance. One explanation for this is the way rural regimes are organised and how they influence regional co-operation.

The regime concept is the subject of a broad debate and is applied in various lines of research (see Horlings *et al.*, 2006 for an overview), including research into policy arrangements (Arts and Leroy, 2006; Van Tatenhove *et al.*, 2000), urban political studies (Stoker, 1995, Stone, 1993) and innovation and transition studies (Geels, 2004; Rip and Kemp, 1998). In the literature on regimes a distinction is also made between sociotechnical systems (Rip and Kemp, 1998) and societal systems (i.e. sectors or regional entities (Rotmans, 2003). The basic assumption in the literature is that a regime is a configuration of political and societal coalitions and institutions, including their discourses and practices, that structure specific parts of society.

The regime concept used in this book is based on the theoretical background formed by urban regime theory in which the regime concept indicates ways in which configurations of private-private, public-public and public-private coalitions are formed by, and in turn structure, urban development agendas (Davies, 2002, 2003; Dowding, 2001; Goodwin and Painter, 1997; Lauria, 1997; Stone, 2002). The concept of regime can be used on different spatial levels. We introduce the *regional regime*, which can be defined, based on Stone (1989) and others, as 'the informal arrangements by which autonomous or semiautonomous actors function together to make and carry out governing decisions relevant for a specific region'. The roots and theoretical background of urban regime theory will be explained in Chapter 3.

According to Stone, one of the founders of this theory, urban regime theory is especially useful in complex situations where there is either little or no hierarchical 'command and control power'. The theory admits the influence of structural factors in the institutional multilevel context, but also analyses the ways actors can change 'the rules of the game'. Until now urban regime theory in the Netherlands has been applied in a local, urban research context (Tops, 2007; Tops and Hendriks, 2004). Our hypothesis is that the theory is open and flexible and suitable for a regional (rural) development context (Horlings *et al.*, 2009b).

Urban regime theory is not only a helpful analytical framework for understanding the problems described in the previous sections, but it can also potentially offer a new perspective on regional complexity and co-operation, as we will see in the following chapters. The regime concept stresses the importance of the mutual embeddedness or co-evolution of governmental and non-governmental/market coalitions. Both in the field of international relations theory and in the field of regional and urban studies, the capacity to act very much depends on co-production or co-evolution, which bring together public and private resources and distribute selective incentives (new jobs, new facilities, new sources of profit, administrative power, etc.). Regimes have histories of collaboration, based not so much on complete forms of consensus over values and beliefs, but rather on a common participation in the realisation of 'small opportunities' (Stone, 1993).

2.6 Problems with public-private co-operation in regimes

With the regime concept we can analyse the problems encountered in region-oriented policy and self-governance described above. Existing regional regimes are hampering regional co-operation in projects based on region-oriented policy and self-governance. Table 2.1 gives an overview of regime problems related to regional development. We identify two main aspects (Horlings *et al.*, 2009b):
1. the predefinition of problems by existing rural regimes;
2. the structure of existing rural regimes and the way participation and co-operation is organised.

2.6.1 Predefinition of problems

To a certain extent, discussions about rural problems are predefined within rural regimes. First, differing views on regional identity can hamper co-operation in regional policy processes. Regional identities are social constructs related to social relationships and can therefore be seen as 'acts of power' (Simon, 2004). Governments often have more power to promote and 'reconstruct' identities than the public. As a result, the definition of rural identities in region-oriented policy is often dominated by governmental institutions. This is particularly evident in the restructuring areas in the Netherlands, where the physical-spatial perspective clashes with the logic of the people. The restructuring plans disregard non-spatial aspects 'that cannot be shown on a map'. People's experiences and inhabitants' sociocultural wishes and non-spatial issues are too often neglected (Haarmann, 2004). Second, different actors often represent different views on sustainable rural development. These different views are not represented equally and have differing degrees of 'persuasive power' in rural development processes, which leads to conflicts and thus hampers co-operation. For example, conflicting discourses arose during the foot and mouth epidemic in the Netherlands, when the image of rurality held

Table 2.1. Rural regimes and problems with co-operation between public and private actors.

Problem	Examples
Predefinition of problems	Power-driven view of identity
	Clashes between discourses
Structure of regimes	Inclusion and exclusion of actors
	Lack of trust
	Incorporation of self-organising initiatives

by farmers, citizens and hobby farmers conflicted with the dominant economic and export-oriented interests of central government (Van der Ziel, 2003).

2.6.2 Structure of regimes

Besides the way in which problems are defined within rural regimes, the process of establishing networks can also hamper co-operation. The main question is: who is in and who is out? The main obstacles to co-operation are:
- inclusion and exclusion of actors in the policy arena;
- creation of distrust;
- incorporation of self-governance groups.

Inclusion and exclusion

Despite the diminishing role of traditional interest groups, new actors face difficulties in trying to enter the national policy arena when new rural policy is being formulated (see for example Horlings, 2004). The question of who can enter the policy arena is power driven. For example, a study on the role of women in restructuring processes in two Dutch provinces showed that, on a regional scale, women play a fairly limited role in the decision processes (Bock *et al.*, 2004).

Lack of trust

Co-operation within rural regimes depends on trust between public and private actors. However, Van der Ploeg describes the community of parties involved in rural development as one that is composed of 'distrust'. Innovative farmers choose their own development path, thus clashing with the development project of the expert system, governments and their own union (Van der Ploeg, 1999). Symptoms of a lack of trust on the part of the government are a focus on the formalisation of agreements, bureaucratisation and an increase in the regulation of regional policy. Explicit and implicit procedures and rules hamper co-operation between public and private parties, and subsidy regulations make it difficult for private actors to obtain resources because they often cannot meet the requirements for co-funding or lack the knowledge and professionalism to make the necessary arrangements (see for example Horlings and Mansfeld, 2006).

Incorporation of self-governance groups

Bottom-up initiatives co-operate with governments in order to gain support or room for manoeuvre. However, they also face the risk of incorporation into institutional structures. On the bases of experiences in Denmark, Bang (2004) sees the threat of an incorporation of conventional practices into the domain of strategic communication between political authorities in the regime. This form of governance is described as an all-pervasive form of colonisation of the lifeworld by systems (Bang, 2004). This leads to a major dilemma for private actors: to get more room for manoeuvre they have to become incorporated into institutional structures, but this can involve high 'transaction costs' and the associated risk of losing vitality.

2.7 Can vital coalitions create capacity to act in regional development processes?

In this section we deal with the question of whether vital interaction can provide a solution to the obstacles in regional co-operation. Interaction with citizens and social and private organisations is an essential feature of modern governance. The emergence of new regimes is one expression of the need to organise public decision-making in an interactive way. But in what situations is co-operation between actors from the public and private sector and civil society desirable, and what is the effective balance between co-operation and autonomy? Some models in organisation theory are available to deal with these questions. In the alliance model public and private actors are equal. In the concession model government decides the conditions for co-operation and which private actors are invited to work together. It is important that in an early stage of a process actors agree on the chosen model and the procedures to be followed. In regional development two forms of alliances can be distinguished:
1. endogenous - initiated by governments, the market or civil society within the region;
2. exogenous - initiated outside the region.

The initiating actor and the phase of the regional process are both relevant. In the implementation phase a specific form of alliance, such as a public private partnership (PPP) or joint venture. A PPP is a form of co-operation in which public and private actors, while maintaining their own identity and responsibilities, realise a project based on a clear division of tasks and risks (Wolting, 2006: 14). The way public and private actors co-operate depends on the issue at stake. Joint responsibilities are feasible in the area of socioeconomic problems in rural areas, for example where people are directly concerned about public interests, such as recreational and environmental issues (WRR, 2006).

Interactivity in regional processes, however, has to be linked with vitality if it is to attract citizens. Modern citizens are not prepared to invest a great deal of time in public decision making if it is not made attractive to them. A link with tangible results and a context that is inspiring and energising is indispensable. In urban regime theory, productive interaction therefore needs to take the form of vital interaction. Following this line of thought, the regional regime approach draws specific attention to new alliances, networks and partnerships based on the uniqueness and qualities of the region (Hamilton, 2004; Horlings *et al.*, 2006). Vital interaction is a form of decision making in which the co-operation between actors is both energising and productive. By productive we mean that results are achieved; by energising we mean that the interactions give energy and inspiration to those who are involved. In an ideal situation, vital interaction takes shape in the form of vital coalitions.

A vital coalition is defined here as 'a form of active citizenship and self-organisation, in which citizens and/or private or public actors take the initiative to act on behalf of a common concern or interest'. The background to this concept is further elaborated in Chapter 3. The hypothesis is that certain combinations of influential actors under certain conditions can organise capacity to act and can change the status quo. Vital coalitions may potentially even change 'the rules of the game', affecting the institutional setting (regime) of regional development. Vital coalitions start out as relatively small initiatives, but can exert multiple effects, making people active, bringing together institutions that have never co-operated before, or mobilising energy within a neighbourhood.

Vital coalitions differ from interactive policy in three main ways. First, private actors or actors in the civil society have the initiative and take responsibility for public goals. Second, the focus is more on informal negotiations, dialogue and personal contacts than on formal interaction, with the aim of creating productive action and capacity to act. Third, the goal is not so much the making of plans or policy, but the development of investment propositions, regional storylines or implementation strategies. Regional storylines can attract people to support a joint agenda, as Stone (1989) has shown in Atlanta (see Chapter 3).

What is the sense of urgency underlying the creation of new capacity to act? As we explained above, while the current arrangements used in Dutch rural policy strengthen governmental legitimacy, the new forms of governance do not effectively address the problem of the 'policy management gap' at the regional level. In itself, the shift from government to governance can be considered to be a positive development. Nevertheless, this process is hampered by clashing discourses, insufficient use of self-regulation and inflexible arrangements. These problems cannot be explained from the perspective of networks alone, but require a regime approach and a perspective on creating enabling power.

Urban regime theory sheds more light on the need for enabling power. It takes account of structural influences on urban regimes, but also stresses that local actors can change 'the rules of the game' through their meaningful conduct. Governmental actors cannot create the necessary capacity to act on their own. To be effective, governments must blend their capacities with those of various non-governmental actors (Stoker, 1995: 58). Urban regime theory explains that the capacity to act is not a given, but must be created and maintained actively. The question is not 'who governs', but how to develop the capacity to govern. Leadership is essential in this process and requires more than just having a formal position of authority (Stone, 1989: 229).

This attention to leadership is not a plea for 'strong leaders', nor merely a question of position, status or formal power. Leadership tasks very much depend on the situation, conditions and circumstances. Especially in regional development, leadership is not a straightforward matter of leaders and followers, but a collaborative process. Sotarauta (2005) refers to this process as 'shared leadership'. How leaders contribute to vital coalitions will be explored in Chapter 5.

2.8 Creating sustainability in the interaction between regimes and vital coalitions

As far as sustainability is concerned, the strategic question is whether vital coalitions can contribute to sustainable development. Sustainability is not in itself a characteristic of vital coalitions; this depends on the goals and actions of the coalition. Vital coalitions are also *not* a prescribed regional transition model, but may be the outcome of various regional processes, such as a regional dialogue, the development of new product/market combinations, organising partnerships, or building urban-rural networks, as we will see later in this book. An important condition for creating capacity to act in these regional processes is 'unfreezing' the institutional, organisational and individual aspects (both structure and attitudes) of current regimes.

Interestingly, as Mossberger and Stoker (2001) show, regimes are not the same as governments. As we explain in Chapter 3, regimes are groups of influential actors who to a large extent determine the decision making. They are not bound to periods of office like governments often are, but exist for longer periods. We see them as coalitions of actors that have built relationships based on trust and security and that are highly interdependent for resources (capital, knowledge, rules). They are the people and organisations that are always invited to participate in formal and informal arenas and processes, thereby reproducing ways of thinking and reaffirming their influence.

New coalitions interact with existing regimes. Regimes carry dominant thinking patterns that are so powerful that all actions seem to reaffirm their existence. They are the source of powerful authoritative and allocative rules and can therefore stimulate or restrict the possibilities of self-organised groups of agents that, like regimes, strive to achieve their goals. One way in which vital coalitions can come about is through innovative socioeconomic opportunities. The driving force for coalition building here are the combined investment opportunities for entrepreneurs who organise themselves around these opportunities in chains, clusters, or regional multifunctional business communities. Potentially promising examples can be found in a number of regions under urban influences, such as Heuvelland in the Netherlands, described by Van der Stoep in Chapter 6 (see also Anonymous, 2005; Horlings and Haarmann, 2007), the region near Cork in Ireland (Sonneveld, 2006), and the South Downs in southeast England (Curré, 2008).

Vital coalitions can be seen as self-organised networks of actors that occur in *niches* (incubation rooms for innovation), a term derived from the concept of sociotechnical regimes used in innovation literature (Geels, 2004). This regime concept is different from the concept used in urban regime theory and is used in the context of transition management. Geels distinguishes different levels within sociotechnical systems: niches, regimes and sociotechnical landscapes (Geels, 2004). The niche level refers to local practice in which actors develop new ideas, or novelties, and new sociomaterial configurations (products, practices, concepts, organisation forms, etc). The regime level acts as a sort of mediator of change. A regime refers to dominant practices, rules and shared assumptions. It is characterised by reconfirmation of existing technologies and strategies and is not inclined to promote change. However, these dominant ways of thinking at the regime level can be changed if innovations hold their ground, evolve into a stable design, become institutionalised and are increasingly adopted by others. The regime level can then play an enabling role, using capital and regulations. In transition theory this breakthrough at the regime level marks the take-off phase of transitions (Rotmans *et al.*, 2001). Actors at the regime level are more inclined to react positively to ideas and innovations from niches when they are in line with gradual social trends. Gradual social trends are part of the third level, the 'sociotechnical landscape', where political culture, social values, world views and paradigms are represented.

Although we will not go into this regime concept in this book, the idea of vital coalitions as a form of niche innovation helps with understanding the relation between self-organising initiatives and existing regimes. In sociotechnical regimes, though, regime influence is mostly regarded as a constraint on innovation, whereas in urban regime theory regimes can also have a positive influence. Of course, we should not forget that regimes consist of actors, whose behaviour constitutes a structure, which in turn provides the ground rules for the actions of and interactions between stakeholders at both the regime and the niche level.

The focus of our research is on the concrete relations between coalitions and regimes which consist of actors that try to organise a capacity to act. In our view, capacity to act can be analysed using the concept of power. Power can relate both to actors achieving goals at the cost of other actors and to collective action in which power is manifested through collaboration. Arts and Van Tatenhove (2004) argue that power can be manifested both through discourses and through organisational aspects because policy agents exert influence not only through organisational resources like money, personnel and tactics, but also through arguments and persuasion. They define power as 'the organisational and discursive capacity of agencies, either in competition with one another or jointly, to achieve outcomes in social practices, a capacity which is however co-determined by the structural power of those social institutions in which these agencies are embedded' (Arts and Van Tatenhove, 2004: 347). The most important of these dimensions seems to be the power exercised through discourse, but other forms of power – authoritative and allocative flows of power – should not be ruled out.

For our research, it is interesting to see how the episodic power of self-organised initiatives challenges these structural forces. Network power or intransitive power (Arts and Van Tatenhove, 2004) can be the determining factor in cases where these initiatives are taken up in practice and supported by regimes. Discovering how connections are made between initiatives and regimes will provide insight into how capacity to act and achieving outcomes is organised.

2.9 Summary and conclusions

In this chapter we have proposed that innovative perspectives on rural development require new forms of co-operation between the state, business and civil society. We evaluated two forms of policy implemented in the last few decades in the Netherlands: interactive region-oriented policy and self-governance. A conclusion is that within these forms of policy the search for innovative new perspectives has been organised within traditional arrangements. The constellation of actors in rural regimes influences the way rural problems are defined. In other words, a regime is a 'value-inclusive system'. New actors, such as non-agricultural entrepreneurs, citizens and urban representatives, face difficulties when trying to introduce their own problem definitions, participate in rural decision processes and co-operate with institutionalised governments. Their room for manoeuvre is limited. A hindering regime factor in regional transition processes is the clear divide that still exists between urban and rural governments and between vertically organised policy domains.

In this chapter it is suggested that regime theory can provide a useful framework for explaining the regional problems described. Another interesting point which emerges is that regimes are described as providing a common basis for co-operation in the form of new *vital coalitions* between actors that can then create a *capacity to act*. At the project level, the concept of vital coalitions is useful for analysing regional processes.

We elaborated further on the mechanisms of agents' capacity to act, based on urban regime theory and the notion of niches in innovation theory. The relation between niches and regimes can be framed by questions about how actors in niches interact with influential actors that belong to regimes in order to get their ideas and initiatives implemented. A more concrete picture can be obtained by studying the relations between niches and regimes that consist of actors that somehow try to organise capacity to act. Our conclusion is that capacity to act can be analysed using the concept of power, defined above as 'the capacity to achieve outcomes'. An important way in which power is exercised is through discourse.

Regimes are not the same as structure or government. Regimes are groups of influential actors, which can be both private and public actors, that to a large extent determine decision making. They are not bound to periods in office like governments often are, but exist for longer periods. We see them as coalitions of actors that have built relationships based on trust and security and that are highly interdependent for resources (capital, knowledge, rules).

In order to realise sustainable development it is useful to see how the episodic power of self-organised initiatives challenges these structural forces. Discovering how connections are made between initiatives in niches and regimes will provide insight into how capacity to act and achieving outcomes is organised in regional development.

References

Arts, B. and J. Van Tatenhove, 2004. Policy and Power. A Conceptual framework between the 'old' and 'new' policy idioms. Policy Sciences 37 (3-4): 339-356.

Arts, B. and P. Leroy, 2006. Institutional Dynamics in Environmental Governance. Springer, Dordrecht, the Netherlands.

Bang, H.P., 2004. Cultural governance: governing self-reflexive modernity. Public Administration 82(1): 157-190.

Bauman, Z., 2005. Liquid Life. Polity Press, Cambridge, UK.

Bock, B.B., .P.H.M. Derkzen and S. Joosse, 2004. Leefbaarheid in reconstructiebeleid: een vrouwenzaak? Een vergelijkende studie van het reconstructieproces in Gelderland en Overijssel. Europese Unie, WUR, Wageningen, the Netherlands.

Boonstra, F., 2004. Laveren tussen regio's en regels. Verankering van beleidsarrangementen rond plattelandsontwikkeling in Noordwest Friesland, De Graafschap en Zuidwest Salland. Dissertation, Katholieke Universiteit Nijmegen, Nijmegen, the Netherlands.

Curré, C., 2008. The story behind the story. Report for the project Lifescape-your Landscape, International Examples of sustainable area development. Telos en Provincie Noord-Brabant, 's Hertogenbosch, the Netherlands.

Davies, J.S., 2002. Urban Regime Theory: A Normative-Empirical Critique. Journal of Urban Affairs, (24)1: 1-17.

Davies, J.S., 2003. Partnerships versus Regimes: Why Regime Theory Cannot Explain Urban Coalitions in the UK. Journal of Urban Affairs, 25(3): 253-269.

Dowding, K., 2001. Explaining Urban Regimes. International Journal of Urban and Regional Research, 25(1): 7-19.

Driessen, P.P.J., 1990. Landinrichting gewogen. De plaats van de milieu-, natuur- en landschapsbelangen in het landinrichtingsbeleid: een wetenschappelijke proeve op het gebied van de beleidswetenschappen. Dissertation, Katholieke Universiteit Nijmegen, Nijmegen, the Netherlands.

Geels, F.W., 2004. Understanding system innovations: a critical literature review and a conceptual synthesis. In: B. Elzen, F.W. Geels and K. Green (eds.), System Innovation and the Transition to Sustainability, Edward Elgar Publishing Limited, Cheltenham, UK, pp. 19-47.

Goodwin, M. and J. Painter, 1997. Concrete research, urban regimes and regulation theory. In: M. Lauria (ed.), Reconstructing Regime Theory; regulating urban politics in a global economy. Sage, London, UK, pp. 13-29.

Frouws, J., 1998. The contested redefinition of the countryside: an analysis of rural discourses in the Netherlands. Sociologia Ruralis 38(1): 54-68.

Geels, F.W., 2004. Understanding the dynamics of technological transitions: A co-evolutionary and socio-technical analysis. Research Policy 33(6-7): 897-920.

Frouws, J. and P. Leroy, 2003. Boeren, burgers en buitenlui: over nieuwe coalities en sturingsvormen in het landelijk gebied. Tijdschrift voor Sociaal-wetenschappelijke aspecten van de Landbouw 18(2): 90-102.

Gijsberts, P., 1995. Gebiedsgericht milieubeleid. In: K. Bouwer and P. Leroy (eds.), Milieu en Ruimte, analyse en beleid. Boom, Amsterdam/Meppel, the Netherlands, pp. 164-184.

Haarmann, W., 2004. Tussen droom en daad; de bijdrage van het reconstructieplan aan de duurzame ontwikkeling in De Peel. Telos, Tilburg, the Netherlands.

Hajer, M.A., 1995. The politics of environmental discourse: ecological modernization and the policy process. Oxford University Press. Oxford, UK.

Hajer, M.A., 2006. Doing discourse analysis: coalitions, practices, meaning. In: M. Van den Brink and T. Metze (eds.), Words matter in policy and planning; discourse theory and method in the social sciences. Koninklijk Nederlands Aardrijkskundig Genootschap, Graduate School of Urban and Regional Research, Utrecht, the Netherlands.

Hajer, M. and W. Zonneveld, 2000. Spatial planning in the network society - rethinking the principles of planning in the Netherlands. European Planning Studies 8(3): 337-355.

Hermans, H., I. Horlings, P.J. Beers and J.T. Mommaas, 2010. The contested redefinition of a sustainable countryside; revisiting Frouw's rurality discourses. Sociologia Ruralis 50(1): 46-63.

Hamilton, D., 2004. Developing Regional Regimes: A Comparison of Two Metropolitan Areas Journal of Urban Affairs 26(4): 455-477.

Hermans, F. and J. Dagevos, 2002. De duurzaamheidbalans van Brabant 2002. Brabants Centrum voor Duurzaamheidvraagstukken, Telos, Tilburg, the Netherlands.

Horlings, L.G., 1996. Duurzaam boeren met beleid; innovatiegroepen in de Nederlandse landbouw. Dissertation, Katholieke Universiteit Nijmegen, Nijmegen, the Netherlands.

Horlings, I. and M. Van Mansfeld, 2006. Back to BSIK? Positionpaper Innovatieproject Green Valley. Telos and TransForum, Tilburg, the Netherlands.

Horlings, I. and W. Haarmann, 2007. The soft stuff is the hard stuff; vital coalitions in rural-urban regions. Paper for the ESRS conference, Hungary.

Horlings, I., P. Tops, J. Ostaaijen and E. Cornelissen, 2006. The urban regime theory as theoretical framework for analysing public-private partnerships and self-governance in rural regions. In: H. Van Latesteijn and H. Mommaas (eds.), The organisation of innovation and transition. TransForum, Zoetermeer, the Netherlands, pp. 3-36.

Horlings, I., G. Remmers and T. Duffhues, 2009a. Bezieling, de X-factor in gebiedsontwikkeling. Telos, Tilburg, the Netherlands.

Horlings, I., P. Tops and J. Van Ostaaijen, 2009b. Regimes and vital coalitions in rural-urban regions in the Netherlands. In: K. Andersson, M. Lehtola, E. Eklund, P. Salmi (eds.), Beyond the Rural-Urban Divide: Cross-Continental Perspectives on the Differentiated Countryside and its Regulation. Research in Rural Sociology and Development, Volume 14. Emerald Group Publishing Limited, Bingley, UK pp. 191-220.

Jessop, B., 1997. The Entrepreneurial City: Re-imaging localities, redesigning economic governance, or restructuring capital? In: N. Jewson and S. MacGregor (eds.), Transforming Cities. Contested Governance and New Spatial Divisions. Routledge, London:, UK, pp. 28-41.

Lauria, M., 1997. Reconstructing Urban Regime Theory: Regulating Urban Politics in a Global Economy. Thousand Oaks: Sage Publications, London, UK.

Lovan, W.R., M. Murray and R. Shaffer, 2004. Participatory governance, planning, conflict and public decision-making in civil society. Ashgate, Surry, UK.

Marsden, T., 1998. New Rural Territories: Regulating the Differentiated Rural Spaces. Journal of Rural Studies, 14(1): 107-117.

Marsden, T., 2003. The Condition of Rural Sustainability. Royal Van Gorcum, Assen, the Netherlands.

Mossberger, K. and Stoker, G., 2001. The Evolution of Urban Regime Theory: The Challenge of Conceptualization. Urban Affairs Review 36(6): 810-835.

Melman, D., 2003. Co-referaat Westhofflezing. Natura, Haarlem, the Netherlands.

Ministerie van VROM, 1989. Vierde Nota ruimtelijke ordening. SDU, Den Haag, the Netherlands.

Ministerie van VROM, 1998. Evaluatie ROM-aanpak, eindrapportage. SDU, Den Haag, the Netherlands.

Mommaas, H., M. Van den Heuvel and W. Knulst, 2000. De vrijetijdsindustrie in stad en land: een studie naar de markt van belevenissen. SDU, Den Haag, the Netherlands.

Mommaas, H. and L. Boelens, 2006. Voorbij het plan: de actorbenadering. In: N. Aarts, R. During and P. Van der Jagt (eds.), Te Koop en andere ideeën over de inrichting van Nederland. Wageningen University and Research Centre, Wageningen, the Netherlands, pp. 153-161.

Mommaas, H. and J. Janssen, 2008. Towards a synergy between 'content' and 'process' in Dutch spatial planning: The Heuvelland case. Journal of Housing and the Built Environment 23(1): 21-35.

Novioconsult, 1998. Eindrapportage evaluatie ROM-aanpak. Novioconsult, Nijmegen, the Netherlands.

OECD (Organisation for Economic Cooperation and Development), 2006. The New Rural Paradigm: Policies and Governance. OECD Publishing, Paris, France.

Oerlemans, N., E. Van Well and J.A. Guldemond, 2004. Agrarische natuurverenigingen aan de slag. Achtergronddocument bij Natuurbalans 2004. WUR, Wageningen, the Netherlands.

Pellizoni, L., 2004. Responsibility and environmental governance. Environmental Politics 13(3): 541-565.

Pleijte, M., R.P. Kranendonk, F. Langers and Y. Hoogeveen, 2000. WCL's ingekleurd. Monitoring en evaluatie van het beleid voor Waardevolle Cultuurlandschappen. Alterra, Wageningen, the Netherlands.

Pike, A., J. Tomaney and A. Rodriguez-Pose, 2006. Local and regional development. Routledge, London, UK.

Rhodes, R., 1997. Understanding Governance: Policy Networks, Governance, Reflexivity And Accountability. Open University Press, Buckingham, UK.

Pröpper I. and D. Steenbeek, 2001. De aanpak van interactief beleid: elke situatie is anders. Coutinho, Bussum, the Netherlands.

Rip, A. and R. Kemp, 1998. Technological change. In: S. Rayner and E.L. Malone (eds.), Human Choice and Climate change. Vol. 2, Resources and Technology. Battelle Press, Columbus, OH, USA, pp. 327-399.

Ritsema van Eck, J. and H. Farjon, 2008. Monitor Nota Ruimte, de eerste vervolgmeting. Milieu- en Natuurplanbureau, Bilthoven, Ruimtelijk Planbureau, Den Haag. NAI uitgevers, Rotterdam, the Netherlands.

RIVM (Het Rijksinstituut voor Volksgezondheid en Milieu), 2002. Natuurbalans. Kluwer, Alphen aan de Rijn.

Rotmans, J., J. Grosskurth, M. Van Asselt and D. Loorbach, 2001. Duurzame ontwikkeling, van concept naar uitvoering. ICIS, Universiteit van Maastricht, Maastricht, the Netherlands.

Rotmans, J., 2003. Transitiemanagement; sleutel voor een duurzame samenleving. Van Gorcum, Assen, the Netherlands.

Schmitter, P., 2002. Participation in Governance Arrangements: is there any Reason to Expect it Will Achieve Sustainable and Innovative Policies in a Multi-Level Context? In: J.R. Grote and B. Gbikpi (eds.), Participatory Governance: Political and Societal Implications. Leske and Budrich, Opladen, pp. 51-69.

Simon, C., 2004. Ruimte voor identiteit; De productie en reproductie van streekidentiteiten in Nederland. Dissertation, Rijksuniversiteit Groningen, Groningen, the Netherlands.

Sotarauta, M., 2005. Shared Leadership and Dynamic Capabilities in Regional Development. In: I. Sagan and H. Halkier (eds.), Regionalism Contested; Institution, Society, Governance. Ashgate Publishing Limited, Hants, UK, pp. 53-72.

Stoker G., 1995. Regime theory and urban politics. In: D. Judge, G. Stoker and H. Wolman (eds.), Theories of Urban Politics. Sage Publications Limited, London, UK, pp. 54-74.

Stoker, G., 1998. Governance as Theory: Five Propositions. International Social Science Journal 50(155): 17-28.

Stoker, G., 2003. Public Value Management and Network Governance: a New Resolution of the Democracy/Efficiency Tradeoff. Draft paper, University of Manchester, Manchester, UK.

Stone, C.N., 1989. Regime Politics: Governing Atlanta, 1946-1988. University Press, Lawrence, KS, USA.

Stone, C.N., 1993. Urban regimes and the capacity to govern. Journal of Urban Affairs 15(1): 1-28.

Stone, C.N., 2002. Urban Regimes and Problems of Local Democracy. Paper prepared for Workshop 6, Institutional Innovations in Local Democracy, ECPR joint sessions, Turin, Italy.

Sonneveld, M., 2006. Gebiedsontwikkeling met de Fuchsiabrand in West Cork. Rapport van de kennisexpeditie vanuit Brabant naar West Cork, Ierland, 3-5 juni 2009, Regiowaarde, Tilburg, the Netherlands.

Torfing, J., E. Sørensen, L.P. Christensen, L.P. (eds.), 2003. Nine competing definitions of governance, governance networks and meta-governance. Working Paper 2003: 1, Centre for Democratic Network Governance, Roskilde University, Roskilde, Denmark.

Tops, P., 2007. Regime-verandering in Rotterdam. Hoe een stadsbestuur zichzelf opnieuw uitvond. Uitgeverij Atlas. Amsterdam, the Netherlands.

Tops, P. and F. Hendriks, 2004. Governance as Vital Interaction. Dealing with Ambiguity in Interactive Decisionmaking. Paper presented at the International Conference on Democratic Network Governance, 21-22 October, Copenhagen, Denmark.

Van den Brink, A., A. Van der Valk and T. Van Dijk, 2006. Planning and the Challenges of the Metropolitan Landscape: Innovation in the Netherlands. International Planning Studies 11 (3-4): 145-163.

Van der Ploeg, J.D., 1999. De virtuele boer. Van Gorcum, Assen, the Netherlands.

Van der Ploeg, J.D., 1992. The reconstruction of locality: technology and labour in modern agriculture. In: T. Marsden, P. Lowe and S. Whatmore (eds.), Labour and locality; uneven development and the rural labour process. Critical perspectives on rural change. Series 4. David Fulton Publishers, London, UK, pp. 19-43.

Van der Ploeg, J.D., A. Long and J. Banks, 2002. Rural Development: The State of the Art. In: J.D. Van der Ploeg, A. Long and J. Banks (eds.), Living Countrysides: Rural Development in Processes in Europe: The State of the Art. Elsevier, Doetinchem, the Netherlands, pp. 8-17.

Van Stokkom, B., 2006. Rituelen van beraadslaging, reflecties over burgerraad en burgerbestuur. University Press, Amsterdam, the Netherlands.

Van Tatenhove, J., B. Arts and P. Leroy, 2000. The institutionalisation of environmental politics. In J. Van Tatenhove, B. Arts and P. Leroy (eds.), Political modernisation and the Environment, the Renewal of Environmental Policy Arrangements. Academic Publishers, Dordrecht /Boston/London, pp. 350-52.

Van der Ziel, T., 2003. Verzet en verlangen. De constructie van nieuwe ruraliteiten rond de MKZ-crisis en de trek naar het platteland. Dissertation, Wageningen University and Researchcentre, Wageningen, the Netherlands.

Verwijnen, J. and Lehtovuori, P. (eds.), 1999. Creative Cities: Cultural Industries, Urban Development and the Information Society. University of Art and Design, Helsinki, Finland.

VROM-raad, 2004. Meerwerk; advies over de landbouw en het landelijk gebied in ruimtelijk perspectief. VROM-raad advies 042, SDU, Den Haag, the Netherlands.

Wiskerke, J.S.C., 2007. Robuuste regio's: dynamiek, samenhang en diversiteit in het metropolitane landschap. Inaugurale rede, 15-11-2007, Wageningen University and Researchcentre, Wageningen, the Netherlands.

Wiskerke, J.S.C. and D. Roep, 2007. Constructing a Sustainable Pork Supply Chain: A Case of Techno-institutional Innovation. Journal of Environmental Policy and Planning 9(1): 53-74.

WCED, World Commission on Environment and Development, 1987. Our common future. Oxford University Press, Oxford, UK.

Wolting, B., 2006. PPS en gebiedsontwikkeling. SDU, Den Haag, the Netherlands.

WRR (Wetenschappelijke Raad voor het Regeringsbeleid), 2006. De lerende overheid. Een pleidooi voor probleemgerichte politiek. Rapport no. 75. SDU, Den Haag, the Netherlands.

ZKA Leisure consultants and planners, Urban Unlimited urban and regional planners, University of Tilburg department of leisure studies, 2005. Heerlijkheid Heuvelland; nieuwe markten en allianties voor toerisme in het Heuvelland. Authorised by LIOF: Maastricht.

Chapter 3
New concepts for studying regional development

Julien van Ostaaijen

3.1 Introduction

In recent years interest in regional and urban-rural developments has grown, both in research and politics (Saartenoja, 2003), and the European Union is putting increasing weight behind co-operation between urban and rural areas. The Leipzig Charter on Sustainable European Cities states that 'coordination at local and city-regional level should be strengthened. An equal partnership between cities and rural areas…is the aim' (Leipzig Charter, 2007: 3). This growing interest raises the need for conceptual tools to analyse these emerging forms of co-operation. We believe that the concepts of urban regime and vital coalitions have the potential to do this. The urban regime has been the 'dominant analytical framework of urban political research in the United States for the past two decades' (Pierre, 2005: 449) and has proven beneficial in explaining local co-operation (e.g. Stone, 1989). Some scholars have also made an attempt to use the concept to analyse co-operation at the regional level. The concept of a vital coalition, on the other hand, focuses more on small, usually private, bottom-up initiatives. We believe the interaction between these two concepts can be fruitful for the regional level: a *regional regime* concept might deliver insight into what hinders or stimulates bottom-up initiatives, and the *vital coalition* concept might help to explain those initiatives. This leads to the following central question for this chapter:

> *How can the concepts of urban regimes and vital coalitions be used to study public-private initiatives in a regional setting, how do these two concepts relate to each other, and what are conditions for vital coalitions?*

This chapter consists of four parts. The first part presents the policy network concept as an overall framework within which the concepts of urban/regional regimes and vital coalitions are compared. The next two parts describe each in turn. The relationship between the regional regime and vital coalition concepts is explored in the last part.

3.2 Urban regimes and vital coalitions: forms of policy networks

The concepts of urban regime and vital coalition may be less well known than the policy network concept, especially among European scholars working in the field of political or related sciences. The three concepts are nevertheless related and display similarities. Within the broad array of literature on policy networks, urban regimes and vital coalitions can even be considered to be specific types of policy networks. This section briefly addresses that similarity and its relevance.

Work on policy networks arose in the social sciences in the 1970s and 1980s and has a strong European background, with Dutch and German scholars contributing much to its development (Bogason and Toonen, 1998: 209; Kickert *et al.*, 1997; Koppenjan *et al.*, 1993). This work reflects experiences with government from the 1960s and 1970s, when non-governmental actors came increasingly into play and the realisation grew that the ability of the government to unilaterally steer society is limited and that it depends on civil organisations and pressure groups to implement policy. As these organisations have a certain degree of autonomy, government authorities cannot adopt a hierarchical position, but must interact with these actors. Interaction can lead to more or less stable patterns around a certain policy area: a policy network (De Bruijn and Ringeling, 1993: 11). According to De Bruijn, policy networks are considered to be 'patterns of interaction between mutual independent actors who form around policy problems or policy programs' (De Bruijn and Ringeling, 1993: 19). Börzel gives the following minimal definition of a policy network: 'a set of relatively stable relationships which are of non-hierarchical and interdependent nature linking a variety of actors, who share common interests with regard to a policy and who exchange resources to pursue these shared interests acknowledging that co-operation is the best way to achieve common goals' (Börzel, 1998: 254).

Urban regimes and vital coalitions share some characteristics with the policy network, such as co-operation between several stakeholders, a common goal or problem interpretation (or at least problem recognition), and mutual dependence. Both regimes and vital coalitions are informal gatherings of semiautonomous actors or individuals that come together for a common purpose and use resources to achieve it. But there are also important differences. Regimes are much more extensive and established than vital coalitions.

> *Where the regime concept is useful on the level of systems, representing a specific integration and co-evolution of the public and private field, the concept of vital coalitions is useful on the concrete level of actors, projects and networks, representing a specific energising and productive collaboration between public and private partners, able to create a 'capacity to act'*
> (Horlings and Haarmann 2007).

In the Section 3.3 we explore the concept of urban regimes and whether it is possible to expand the concept to the regional level.

3.3 The urban regime concept

3.3.1 Stone's Atlanta urban regime

In 1989 Clarence Stone's *Regime politics governing Atlanta 1946-1988* was published. In this book Stone uses what he calls an 'urban regime' to describe the government of Atlanta in the US for a period of over forty years. Many consider Stone's book to mark the beginning of the debate about the use and value of the urban regime concept (see for example John, 2001; Lauria, 1997; Mossberger, 2008; Pierre, 2005; Sellers, 2002; Stoker, 1995).

Clarence Stone defines an urban regime as 'the informal arrangements by which public bodies and private interests function together to make and carry out governing decisions' (Stone, 1989). His main question is empirical. Stone wanted to find out why civic life in Atlanta during the decades after the Second World War, unlike other cities in the southern United States and despite the tensions in society, was not dominated by racial polarisation. To answer this question, Stone studied over forty years of municipal politics in Atlanta, from 1946 to 1988. When in 1946 the black population of Atlanta obtained the right to vote, it increasingly became a force to be reckoned with within the city. One of the demands made by the black electorate was for more housing and living space. The elected members of Atlanta City Council became increasingly aware of this and the first few mayors from the 1946-1988 period had to take this new electorate into account, winning their votes by gradually repealing racial regulations. In 1973 the link between city hall and the black electorate became even stronger when the first black mayor was elected.

In *Regime politics*, Stone describes the behind the scenes negotiations between city hall and the business elite, and how this coalition was able to pursue an agenda which benefited both groups. Under the slogan 'the city too busy to hate', working relations were established between the black middle class electorate, increasingly represented by city hall politicians, and the mainly white and organised business elite. At first sight, they seemed to have little in common, but both parties wanted change. The predominantly white business elite, represented in the Central Atlanta Progress, wanted to see Atlanta adjust to the automobile and service era, and business investments and loans to help the city overcome its financial problems. They also wanted to keep black expansion under control, but without racial unrest, as this would damage the image of the city, and therefore business. They lacked the numbers to become a strong electoral force, but they had other resources they could mobilise. The growing, largely middle class black electorate wanted more housing opportunities and removal of obstructions to black participation in public life. Both could only achieve their aims by working together. The business elite possessed the financial resources for investment, while the black middle class electorate provided the political leadership, both indispensable resources for achieving their aims. The behind the scenes negotiations result in a long-lasting co-operation between city hall and the business elite. Stone calls this an urban regime (Stone, 1989).

Regime politics also describes some more theoretical characteristics of an urban regime. First, an urban regime is about the *informal arrangements* that surround the formal working of government. If cities are 'organisations that lack a conjoining structure of command', Stone wants to know 'how in the face of complex and sometimes divisive forces, an effective and durable capacity to govern can be created' (Stone, 1989: xi). The answer is the informal coalition. Regime building is about establishing new ties between organisations, about how different norms and values existing within institutions relate to each other when working together within an urban regime. Trust plays an important part in forming these arrangements. It is important that the main actors or individuals in a regime trust each other and know that they can count on each other. Trust and regimes can only be built through repeated interactions.

The informal group has to be relatively *stable* and have *access to institutional resources* to execute the agenda of the urban regime. Informal co-operation is not a goal in itself. In Atlanta it arose from necessity, as each groups needed the other's resources; the business elite had financial power, the black middle class electoral power. Moreover, the coalition was stable, as Stone emphasised by pointing out that the parties in the regime co-operated for more than forty years. Neither of these two aspects implies that the relationship also has to be equal. In Atlanta the business community was tightly knit and organised, and it exerted more influence over city hall than vice versa, but this did not weaken the mutual dependence. An urban regime is characterised by mutual dependence, not necessarily by equal dependence.

Finally, a regime is *empowering*. A regime wants to get results that cannot be achieved without the co-operation of the partners. In Atlanta the results varied from building an expressway system to developing new housing for blacks, desegregating the school system, and organising a National Black Arts Festival. All activities were the product of co-operation between regime partners.

For Stone, governing in an urban regime is not about command and control. It is about how different actors can get things done. Given the fact that communities are divided into many sectors with independent actors, co-operation cannot be achieved by exerting hierarchical power: *power over*. Co-operation is obtained through informal arrangements between more or less autonomous actors: *power to*. *Power over* can be exercised over some groups, but cannot be used to dominate other independent groups. In Atlanta the business community and city hall are both autonomous groups and have certain powers and capacities, but not *power over* each other. For Stone, *power to* makes his concept of urban regime different from elitism and pluralism, which are more concerned with the 'social control model' of power over, the power to control or dominate others. Gerry Stoker (1995) mentions that *power to*, also called the *power of social production*, is the important form of power in an urban regime. For him, *power to* is about the ability of a group of actors to form a structure that can solve collective action problems, build a regime and gain the capacity to govern. This form of power is intentional and active: it accomplishes, it has the ability to produce results. Stoker distinguishes this form of power from three other types of power. The first is *systemic power*, which has to do with the formal position of actors within the social and economic structure. This position can be so self-evident that actors do not even have to act; their power is a matter of the context, of the logic of the situation. *Command power* is a more active form of power that mobilises resources, such as information, money, knowledge or reputation. This kind of power is only active in limited areas and for certain activities. The third category of power, a response to the limitations of command power, is *coalition power*. It deals with the way in which actors use their relative autonomy when negotiating with other actors to form a coalition (Stoker, 1995:64-65).[1]

3.3.2 Background of the concept

The urban regime does not have its roots only in the policy network concept. Stone also acknowledges the work of other authors in his interpretation of what an urban regime is. He was inspired by Stephen Elkin (1987), who does not use the concept of a regime to mean the same as Stone, but talks about alliances between public officials and business leaders. His case study of Dallas in particular comes close to what Stone would later describe as an urban regime. In fact, other scholars had used the term 'urban regime' prior to Stone's research

---

[1] For more on the concept of power in policy-making processes see, for instance, Arts and van Tatenhove, 2004.

in Atlanta. Fainstein and Fainstein described a local regime as a circle of powerful local government administrators and elected officials who move in and out of office, to distinguish it from the more stable and overarching entity of the state (cited in Mossberger, 2008: 41-42). Reed (1988: 138) described black-led and black-dominated administrations backed by solid council majorities as 'black urban regimes'.

The idea of regimes also appears in other disciplines. In organisational theory regimes are used as a typology to classify different organisations (Lammers, 1983). One distinction is between 'mechanical' and 'organic' regimes. A mechanical regime is characterised by hierarchy and co-ordination from the top, which gives it a specific role in the organisation; an organic regime has a more organic place within the rest of the organisation and adjusts itself through interaction with other parts of the organisation (Lammers, 1983: 147). Regimes also appear in the natural sciences and other disciplines, but the discipline most applicable to the urban regime concept is international relations.

From the international relations regime concept, the idea of mutually beneficial co-operation between relatively autonomous actors has made important inroads into the urban regime literature. Regime theory emerged in the international relations discipline in the 1970s. It reflected the growth of international non-government organisations and increasing co-operation between countries without compulsion by any hegemonic power (Junne, 1992: 9). According to Humphreys (1996), Ruggie was the first to use the term 'international regime', in 1975 when he defined a regime as a set of mutual expectations, rules and regulations, plans, organisational energies and financial commitments which have been accepted by a group of states (Ruggie cited in Humphreys, 1996). A definition that has become more common is that a regime consists of sets of implicit or explicit principles, norms, rules and decision-making procedures around which actors' expectations converge in a given area of international relations (Krasner cited in Humphreys, 1996). Regimes therefore function as a stable set of principles, norms, rules and procedures that guide international co-operation (Junne, 1992: 9).

International regime theory emphasises the autonomy of actors and the need for co-operation to 'get things done'. Although this characteristic is clearly visible in the urban regime concept as well, this concept is not just a transformation of these ideas to an urban level. Stone's urban regime emerged from the desire to answer a specific local question and connects it with certain specific characteristics of urban regimes. This gives the urban regime its own dynamics and meaning.

3.3.3 Urban regime characteristics

Broadly speaking, an urban regime is an informal coalition dealing with internal and external challenges. In the case of Atlanta, these challenges are metropolitan decentralisation, changing race relations and mobile capital. At that time, these challenges may not have been much different from other cities, but, Stone argues, the way Atlanta adjusted to these changes can only be explained by looking at the specific local circumstances.

> *If Atlanta is exceptional, its exceptionalism lies primarily in the strength and ability of its governing coalition to carry out an activist agenda in the face of resistance and opposition.*
> (Stone, 1989: 178)

The development of an urban regime is not always a process in which all the members see this big picture immediately. Some business elite members in Atlanta were in favour of the segregation system and some black middle class members were not always happy with the close connection between city hall and the business elite. The regime formation process is therefore never fixed. It is the interaction between all these participants that gives direction to the regime. Selective incentives are important in the establishment of a regime.

> *Most people most of the time are not guided by a grand vision of how the world might be reformed, but by the pursuit of particular opportunities…perhaps a nonprofit housing venture, a community theatre, a job-training program, saving a black business from financial setback, conservation of park land, a food bank for the hungry, an historic preservation ordinance, or an arts festival.*
> (Stone, 1989: 193)

Establishing and maintaining the biracial coalition in Atlanta was about 'struggle and conflict', in which the regime is always subject to alteration as a result of internal and external conflicts. Internal conflicts are when the partners constituting the regime clash. In Atlanta, Mayor Jackson initially tried to work more independently of the business elite, but soon learned that co-operation was more fruitful ('go along to get along'). External threats are when the urban regime is threatened from outside the regime. Since an important aspect of an urban regime is that it also excludes actors not needed to achieve the aims of the regime, there are always a substantial number of actors that do not profit from the regime building because they are not a part of it, or do not want to be a part of it, or because they do not subscribe to its agenda. When a substantial body of people who believe in a new way of doing things form a partnership that is strong and workable enough, and do not fear the short-term disadvantages (the punishments for 'not going along'), a regime can be overthrown. This is a very difficult process because the opponents must not only deliver new regime partners and new ways of interacting, but also a new set of ideas. Both internal and external conflicts test the strength of the urban regime.

In later work, Stone also points out four basic 'ingredients' or 'elements' of an urban regime:
- an *agenda* to address a distinct set of problems;
- a governing *coalition* formed around the agenda;
- *resources* for the pursuit of the agenda, brought to bear by members of the governing coalition;
- a *scheme of cooperation* (also referred to as a 'mode of alignment') through which the members of the governing coalition align their contribution to the task of governing (Stone, 2005: 329, emphasis added).

In hindsight, Stone applies these elements to his Atlanta research:
- agenda: 'a city too busy to hate'
- governing coalition: black business + business elite
- resources: control of city hall, investment capital, money, civic skills, political access to governor's office
- scheme of cooperation: behind-the-scenes negotiations

Stone explains that the agenda has always been a central aspect in any urban regime, by which he means 'the set of challenges which policy makers accord priority' (Stone, 2005: 1). All the urban regime actors have their own agenda. Agendas compete for dominance, but 'new' overarching agendas do not always emerge. Sometimes one agenda does become dominant and a coalition forms around it. This is a very dynamic process and the agenda and the coalition are in a constant state of flux.

> *Agendas are never static, and they undergo adjustment as conditions change. But the direction of the adjustment is influenced by the particulars of the network, who composes it, and the concerns they embod.*
>
> (Stone, 2005)

In Stone's urban regime concept, the agenda should apply to the whole city and not just a part of it. Stone acknowledges that agendas fight for dominance, but his Atlanta urban regime is built around what eventually became the dominant agenda for the city: a city 'too busy to hate'. It focuses on economic development and rolling back racial exclusion, and therefore gives direction for governing the entire municipality. The agenda is the 'cement' that holds the governing coalition together despite internal differences, especially in the beginning when the participating actors learn about each other and when trust is built. The coalition, according to Stone, is:

> *the group of actors who come together, in many instances unofficially and tacitly, for the purpose of setting a locality-wide agenda and giving it priority standing (that is, they provide 'guiding and steering').*
>
> (Stone, 2004: 3)

To bring together and maintain a governing coalition, the agenda should be appealing. It has to be broad enough to appeal to possible regime partners and to gain citywide priority. On the other hand, an agenda cannot be too broad.

> *If the agenda is too narrow, then it lacks a capacity for direction setting and has a weak claim for priority status…If the agenda is too broad, it may lack focus and lose its staying power.*
>
> (Stone, 2002: 7)

In Atlanta, this was captured in the slogan 'the city too busy to hate'.

> *The agenda has been sufficiently broad, yet focused, to accommodate a long succession of individual projects and initiatives, on which a tradition of biracial cooperation now rests.*
>
> (Stone, 2002: 8)

An urban agenda can encompass a single policy area, after which it broadens to other urban policy areas. New initiatives can emerge in one or more policy areas, allowing it to function also as an umbrella for other projects.

> *The strength of the Atlanta regime lay in its ability to pass on a certain way of working, a set of assumptions, and a shared view of the future of the city. A regime is more than the sum of the individuals involved. It creates a framework of incentives and meaning in which regime partners act and into which new actors can be incorporated.*
>
> (Orr and Stoker, 1994: 67-68)

To accomplish the agenda, actors need to bring in resources. These can be tangible, such as money or material, but also less tangible, such as knowledge or status. The way different actors interact with each other around the agenda is called the mode of alignment or scheme of co-operation.

3.3.4 Main discussion points in the urban regime debate

The urban regime concept quickly became popular in the US and it is a familiar concept in the study of urban processes. Numerous case studies have been done, mainly in the US and to a lesser extent the UK, to determine whether urban regimes exist in places other than Atlanta, and doubts have been raised about use of the concept outside these two countries, or even outside Atlanta. In Europe, for instance, the relationship between local government and the state is said to be different, making the specific circumstances in Atlanta a constraint on wider application of the concept. Three discussion points have emerged in the academic debate:
1. Which actors form the coalition?
2. How durable is an urban regime?
3. How local is, or should, an urban regime be? (See also Van Ostaaijen, 2010)

Actors

The first discussion point is about which actors are needed to constitute an urban regime, and especially the role of business. In Stone's Atlanta, top businessmen functioned as an important urban regime partner. Some authors argue that the combination of business and politics is less relevant for European or other non-American cities. Relations between city governments and central government and the ways in which cities acquire revenues differ a great deal. Most European countries receive much of their income from the state. In the US federal funding is more limited, making it necessary for cities to look for other sources of income, such as the private sector. This increases local dependence on business, but even though businesses in the US are more locally focused than those in Europe, this is not necessarily a reason to change the concept (Mossberger and Stoker, 2001: 819-820). Stoker criticises the example of Menahem

(1994), who describes a partnership without a private sector input, where the major actors were bureaucrats from local and central government. For Stoker, this example stretches the urban regime concept beyond its original meaning (Stoker, 1995). Mossberger and Stoker suggest that in such a case a different concept should be used, for instance, a network (Mossberger and Stoker, 2001: 817). They agree that in different contexts, other actors, such as bureaucrats, neighbourhood councils or civil rights movements, can be part of a regime, but the essential co-operation is between politics and business. Without these, there is no regime (Mossberger and Stoker, 2001: 831).

This strict interpretation can be awkward when applying the concept outside its Atlanta or US context. If the concept is to have value in Europe, or anywhere else, then it is better to talk about co-operation between relative autonomous actors (Van Ostaaijen, 2010). When actors achieve a level of autonomy, it cannot be assumed that they will automatically comply with the urban regime. They have to be persuaded. 'Business' can qualify as a relatively autonomous actor, but so can government and semigovernment actors such as the police, neighbourhood organisations and welfare or health organisations.

Durability

A second discussion point within the urban regime debate is durability. Most authors agree that without a certain amount of durability there is no regime, or at best an emerging regime or a failed regime (Mossberger and Stoker, 2001: 830). Durability also distinguishes urban regimes from other concepts, such as pluralism (Stoker, 1995: 59). Sites, who describes regime change under three consecutive mayors in New York, is criticised for confusing temporarily strategic policy shifts with regime change (Mossberger and Stoker, 2001; Sites, 1997; Stoker, 1995). According to Mossberger and Stoker (2001: 813), it is not exceptional for a regime to span a number of administrations. But it remains less clear what the authors understand by a 'longstanding pattern of co-operation' (Mossberger and Stoker, 2001: 829), which they say must be present. Stone acknowledges that the distinction between the adaptation within a regime and a succession of different regimes is 'inevitably somewhat arbitrary' (Stone, 1989: 181). Stone seems to make the condition of durability concrete by connecting it to the agenda of the regime, emphasising the importance of looking beyond short-term results. An urban regime is a 'set of informal but relatively stable arrangements by which a locality is governed', in which govern should be understood as 'bringing together the resources needed to pursue…a strategic policy direction' (Stone, 2002: 7). It is not complete control over all decisions, but the capacity to give priority to a direction-setting agenda. Stone therefore interprets durability as the capacity of the regime to act out a durable policy agenda.

The length that an urban regime should sustain itself is seldom made concrete (in years, terms, etc.), but Stone's 40-year Atlanta regime always seems to cast a shadow on this debate and tends to be considered the ultimate stable regime. Although it does not seem wise to attach a certain period *in years* to which an urban regime should comply, it is acknowledged that an urban regime should have some aspects of durability. For urban cases, this can be interpreted as meaning more than one legislative period (Van Ostaaijen, 2010).

Local

A third critique of the urban regime concept is that an urban regime is too local, by which it is meant that scholars using urban regimes do not incorporate important non-local actors into the analysis, or that they disregard non-local developments. This makes the local decision-making process more important than it is. Boundaries between states are rapidly disappearing in the globalising economy (see for example Sassen, 1991). If global developments (economy, safety, immigration, etc.) exert such an influence that not even states can control them on their own, then neither can cities. The relative powerlessness of cities had already been shown in Peterson's book *City limits* (Peterson, 1981). Stone and other urban regime scholars do not regard the local level as an inappropriate level of analysis: urban regimes are not isolated and there is always influence from outside the city. In the preface to *Regime politics* (1989), Stone stated that national developments (the abolishment of the Jim Crow system, national urban development programmes) and even global developments (the internationalisation of economies) affected the way the local actors could and did play a part within the urban regime. In fact, the way local actors deal with this is precisely what an urban regime is all about. This does not preclude the local as the starting point of the analysis. In *Regime politics*, for instance, the relationship between city hall and the governor is important. In a later article, Stone describes the multilevel interaction in Boston, a co-operation extending to several federal agencies (Stone, 2005).

But the critique that urban regimes are too local also has another aspect: an urban regime should not be restricted to urban borders. Urban regimes should be able to exceed the 'local' itself, for example by making a regional regime possible.

3.3.5 Towards a regional regime

Hamilton was one of the first to study the application the urban regime concept to regional cases, more specifically the regions of Chicago and Pittsburgh (Hamilton, 2002, 2004). A regional regime goes beyond just the governing of a city or municipality to encompass the interplay between an urban centre and its immediate surrounding. In Hamilton's cases, this mainly means smaller towns and suburbs close to Chicago and Pittsburgh, hence the name regional regime.

Hamilton notes that co-operation on the regional level is much weaker than within many urban regimes. He considers the lack of established patterns of co-operation between regional actors to be an important barrier to the formation of regional regimes, mainly because there is no strong political authority at the regional level (Hamilton, 2002). According to Hamilton, business sometimes takes the lead, but the important political actors are more embedded within their electoral territory, usually the city, and not the regional entity.

As far as concrete results go, Hamilton still feels he only 'scratched the surface' of the application of urban regimes in regional development. He calls for the applicability of the concept to be tested by case studies in other regions (Hamilton, 2004).

3. New concepts for studying regional development

I contend that urban regime theory is an appropriate theory to study and analyze the relationship between politics and economics in regional governance…Applying regime theory to regional governance appears to be a reasonable expansion of the theory.

(Hamilton, 2002: 404, 406)

Using the urban regime concept in the analysis of co-operation between different urban areas is only a small step away from using it to analyse co-operation in regions containing urban and rural areas. Saartenoja for instance talks about 'urban rural regimes' as a specific type of regional regime, focused on urban-rural areas (Saartenoja, 2003). According to Saartenoja an urban rural regime overlaps an urban and rural area and functions more or less in the same way as an urban regime; both are aimed at achieving results (Saartenoja, 2003). Saartenoja acknowledges that the relationship between a politically dominant urban centre and the surrounding rural area is often unbalanced, but this does not make the concept less useful for analysing the interactions that take place between urban and rural areas. Rural development processes often take place in an environment as complex as urban surroundings. The main questions in regional regime analysis encompassing both urban and rural areas are: what is the capacity of urban and rural actors to co-operate, and do they have common aims (Saartenoja, 2003)?

The possibility of overlaps between regional and urban regimes completes the picture. A city can have an urban regime while also being part of a larger, regional regime (Leo cited in Saartenoja, 2003). In other words, a partly overlapping urban regime and regional regime can function at the same time.

Referring to Stone (1989), Saartenoja (2003) and Hamilton (2002, 2004), we now come to a definition of a regional regime:

A regional regime is the informal arrangements by which autonomous and semiautonomous actors work together to make and carry out governing decisions relevant to a specific region.

As Hamilton already noted, urban regime analysis does not differ much when applied to regional or urban rural areas. Regional regimes are also about how a certain agenda takes shape and becomes a priority, how the necessary actors around it are mobilised, what resources they bring to the coalition and how an informal co-ordination mechanism is established. The coalition can consist of governmental, business and/or societal actors. Government does not necessarily have to be the instigator within a regime.

The basics of an urban regime are also relevant when using the regime to study regions:
- an *agenda* to address a distinct set of problems;
- a governing *coalition* formed around the agenda;
- *resources* for the pursuit of the agenda, brought to bear by members of the governing coalition;
- a *mode of alignment* through which the members of the governing coalition align their contribution to the task of governing.

Looking at urban-rural development through the concept of a regional regime can help explain certain peculiarities. Regime analysis highlights the informal aspects of governing alongside the formal working of governing that can be seen within organisations and in the interaction between actors. Regime analysis can shed light on the why some agendas achieve priority and are implemented while others do not. Stone explains this by means of an urban example: Suppose for instance that a city experienced a failed school reform. By putting the circumstances of the city into the model (agenda, coalition, resources, mode of alignment), it may become clear that the city lacked the resources for school reform, the agenda did not become priority, or a mode of alignment between the necessary actors was never established (Stone, 2005: 17). It is also possible to determine the 'winners' and 'losers', usually the actors that gain from a regime and the ones that do not (see for example Burns and Thomas, 2004). A regime always consists of a limited number of participants. Those excluded either oppose the regime or want to be part of it, and are often the ones that least benefit from it. The regime limits participants because having too many actors affects the regime's ability to act, but so does having too few actors. To gain regional priority the urban regime should include powerful actors.

Summary of the regional regime

A regional regime is the informal arrangements by which autonomous and semi-autonomous actors work together to make and carry out governing decisions relevant to a specific region. This means that an agenda should have become a regional priority, a coalition is in place to implement the agenda, there is an informal scheme of co-operation between the coalition partners, and the coalition possesses and uses the necessary resources to implement the agenda. Moreover, a regional regime should have a certain amount of durability, as this is one of the ways in which it is distinguished from the concept of a vital coalition. In the next section, we discuss the latter concept more thoroughly.

3.4 Vital coalitions

The concept of a vital coalition is derived from the work of Hendriks and Tops (2002, 2005), inspired by the work of Bang and Sörensen (1998). Vital coalitions are initiatives by citizens, with some help and support from local government. They are an active form of participation, inspiring the participants to take the initiative. The word 'vital' is derived from the Latin word for life: *vita*.

> *A vital coalition is a form of active citizenship and self-organisation in which citizens and/or private or public actors take the initiative to act on behalf of a common concern or interest.*

The concept of vital coalitions is a reaction to earlier types of citizen participation, such as consultation and interactive policy making. Consultation originates from the 1970s as a right achieved by citizens to have their say in policy projects. Public consultation procedures are now common practice and this right of the citizen has now become a duty for public officials. The same can be seen with interactive policy making, which appeared in the 1990s. Interactive policy making (or co-production) was originally proposed to allow citizens not only to make their views on the policy making process known when the plan is finished (as with consultation), but also to be included in policy-making processes from the beginning (see also De Graaf, 2007). Interactive policy making is a process in which citizens and public officials, often bureaucrats, develop and implement policy. One or two decades later, these forms of policy making face the same fate as consultation decades before: the procedures are being taken over by administrators as a 'normal procedure' in making policy (although not yet at the level of 'consultation', which is even in some cases required by law). This shows the success of the initiative, but when incorporated into standard procedures, interactive policy making does lose some its original vitality. The parameters for the process are then determined by the procedures of bureaucrats, not the wishes of citizens (Boogers et al., 2002a). Public meetings, for example, are held at times and in places decided by government officials. By complying with government procedures, interactive policy making has lost some of it its ability to renew and inspire (Edelenbos and Monnikhof, 2001; Kalk and de Rynck, 2003).

In vital coalitions, on the other hand, citizens, not government, take the initiative. People do not have to be 'dragged to meetings', as in some interactive policy processes (Boogers et al., 2002a, 2002b). Citizens participate in vital coalitions because they feel it will be useful, interesting and necessary. Vital coalitions therefore possess more vitality and consist of people with a willingness to act. They are energising and productive. Interactions between actors, as in interactive policy making, are necessary but not sufficient. All participants within the coalition need to have a certain amount of zeal, activity and vitality. Citizens often take the initiative; they take part on their own conditions and for their own reasons. A coalition is vital if it knows how to act with energy, if it can get things done. The role of local government is also fundamentally different from interactive policy making, shifting from a more steering and directing role to a more supportive and facilitating one. We have seen good examples of vital coalitions in two Rotterdam cases, described in Box 3.1.

The Hand In Hand Hillesluis project was considered to be a success because it was based on a general feeling within the neighbourhood and because it was a citizens' initiative. The members of the vital coalition were not all white and male; they represented the entire neighbourhood. Many people from different ethnic groups were involved and young people even played a part in leading the discussions. The role of the district councillors was important, supporting the initiative, but keeping a low profile. It remained a citizens' initiative. The vital coalition in Pendrecht emerged because there was a shared feeling of discomfort about the way the way the media and politicians were treating the neighbourhood. The idea of putting the biggest Christmas tree in Rotterdam in Pendrecht motivated some neighbourhood citizens to show that Pendrecht was a lively and vital neighbourhood and not just a cluster of houses for poor immigrants. Mobilising other residents and finding sponsors created a broader informal network within Pendrecht of people who supported the initiative. Local government, in this case a district councillor, played an important supporting role in the realisation of this citizens' initiative.

Vital coalitions want to achieve something, which drives the people involved to accomplish things. In Hillesluis it was to reduce the negative tension in the neighbourhood following the murder of a well-known Dutch film and opinion maker in Amsterdam. In Pendrecht, in response to the bad publicity, the goal was to have the highest Christmas tree in Rotterdam in

Julien van Ostaaijen

Box 3.1. Vital coalition: two cases in Rotterdam[1].

Hand In Hand Hillesluis

In early November 2004 there was considerable friction between the various groups of residents in the Hillesluis neighbourhood in Rotterdam, where about three-quarters of the population are immigrants or descendents of immigrants. The district to which Hillesluis belongs (Feijenoord) also has an above average number of young people. The tension in the neighbourhood followed the murder in Amsterdam of Theo van Gogh, a Dutch film director, by a man of Moroccan/Islamic background. When firebombs were thrown at the Breeplein Church in Hillesluis, a number of residents launched a counter campaign with the message: we do not want this in Hillesluis! In a direct reaction to the fire bomb attack, residents had decided to meet in the Breeplein Church to talk about what they themselves could do to coexist peacefully. This first, spontaneous appeal brought together about 200 residents from various ethnic backgrounds and ages at the first Hand In Hand meeting. A second Hand In Hand meeting took place in a Turkish mosque. Once again, almost 200 residents participated, both from within the neighbourhood and beyond. Discussions based on various propositions took place at each table, with themes such as 'What unites us?' and 'Respect'. Fourteen days later, the third Hand In Hand meeting was organised by a Moroccan mosque and held in a community centre, where more than 150 people grouped around tables to talk. Discussions were led by young table chairpersons focusing on the theme 'How do you experience living in the neighbourhood'? Subsequently, a joint festive new year celebration for all residents was organised with district councillors. The Hand In Hand initiators were mainly responsible for the organisational aspects, helped by some young people. During the celebrations, Hand in Hand In Hillesluis was awarded a cheque for 1,500 euros, and the district supported the printing of leaflets, enabling the organisers to continue their activities, including a new Children Hand In Hand initiative at schools.

Vital Pendrecht (*Vitaal Pendrecht*)

Pendrecht is a neighbourhood in the Rotterdam district of Charlois. After the Second World War it housed many dock workers' families. In the last twenty years the neighbourhood has attracted many, often poor, immigrants. The turnover is high and many older residents have moved out. In the summer of 2003, Pendrecht was the focus of a national political and media discussion about whether or not immigrants (or people with a low socioeconomic status) should be spread over the city or clustered.

When a Charlois district councillor heard about the concept of vital coalitions, he asked if this would be possible in his own neighbourhood. Some researchers from Tilburg University then talked to some of the residents, who then discussed what was possible among themselves. During these talks, a few things fell into place. Some of the neighbourhood residents were unhappy about the bad press the neighbourhood was receiving. They had already been active and working hard to improve their neighbourhood, and felt that the bad publicity was ruining all their hard work. The mood was 'we won't let the press destroy our neighbourhood!'

During the residents' meetings the idea came up of having a Christmas tree in the central square that would be one metre higher than the Christmas tree in front of Rotterdam City Hall. This immediately appealed to most residents and three of them arranged it. They succeeded in mobilising many people in the neighbourhood, and finding sponsors did not prove to be a problem. The district councillor supported the initiative, but did not interfere, although on crucial points his help was essential. For instance, when the organisers forgot to apply for a license to erect a Christmas tree in the square, he promised to take care of this, and did. After the tree had been put up, *Vitaal Pendrecht* became an informal organisation and organised other neighbourhood activities in subsequent years (see www.vitaalpendrecht.nl).

---

[1] Examples from Van Ostaaijen and Tops, 2007; the Hillesluis description is derived from Urbact, 2005.

Pendrecht. However, the process is as important as the result. People that share the same goal and enthusiasm are drawn together and reinforce each other's enthusiasm. The co-operation – often with local government – is meaningful in itself. In Pendrecht, the first meeting led to the formation of a group of active citizens within the neighbourhood that, six years later in 2009, still organises activities in the neighbourhood. In Hillesluis the first residents meeting led to a series of meetings about how to live together in the neighbourhood.

Vital coalitions are not limited to the urban context; similar initiatives can arise in urban-rural areas as well. The following example (Box 3.2) from Van der Stoep and Van den Brink (2006) shows signs of a vital coalition like those described in urban contexts.

Even though on a smaller scale, vital coalitions share some regime and network characteristics, such as non-hierarchical linking of a variety of actors and recognition that co-operation is the best way to achieve common goals (Börzel, 1998: 254), they are not 'governing coalitions', they are characterised as 'informal arrangements' that 'carry out…decisions'. Descriptions of the concept of vital coalitions have also made use of the four urban regime elements of agenda, coalition, resources, and mode of alignment (Horlings and Haarmann, 2007, 2008).

Vital coalitions have to contend with different logics (Figure 3.1), which can make co-operation difficult. Two sources of tension are important here: the friction between institutional and situational logic, and that between instrumental and cultural logic. In vital coalitions it is important to deal with these tensions, but they should not be removed entirely because they are necessary for maintaining focus within the coalition (Tops and Hendriks, 2004).

The difference between institutional and situational logic is the difference that is found between universally applicable rules and procedures (often maintained by governmental actors and applied in concrete situations) and reasoning from a concrete situation (often by citizens). The other tension is between cultural and instrumental logic. The instrumental perspective on interaction within vital coalitions is to analyse and solve a certain 'problem' as efficiently and effectively as possible; the cultural perspective rests on the assumption that the process of solving problems is valuable in itself. Some people or actors participate in a vital coalition

---

Box 3.2. Vital coalition: Stolwijkersluis.

In 2006 a plan for the development of the rural-urban area of Stolwijkersluis in Gouda was presented to the responsible councillor. It was prepared by a group of enthusiastic residents of Gouda and the neighbourhood of Stolwijkersluis. With negotiations on the restoration of a sluice in the area stalled, the resulting impasse and lack of decisiveness by government prompted these concerned residents to explore alternatives for the restoration of the sluice. They asked the Land Use Planning Group at Wageningen University to do the study, which a year later was extended into a design process that went beyond the restoration of the sluice and emphasised the social and spatial relations with the surrounding area. At this point, the process facilitators asked representatives from Gouda Municipal Council and the Schieland and Krimpenerwaard water board to participate and finance part of the programme. Seeing opportunities to resolve the conflict about the sluice, which by then had already taken twenty years, they agreed to participate (Van der Stoep and Van den Brink, 2006).

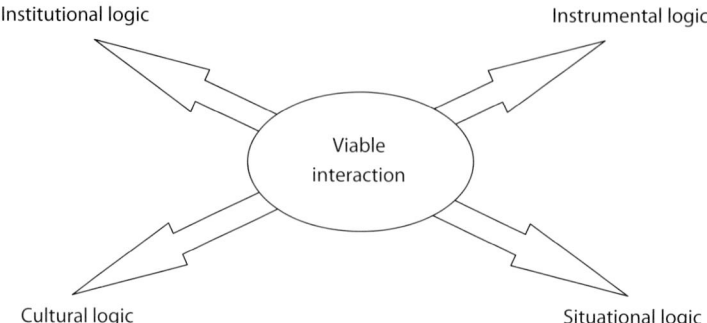

Figure 3.1. Logics within vital coalitions (Tops and Hendriks, 2004).

not only to solve the problem, but also because the co-operation and interaction around the problem holds value for them and gives them energy.

These tensions are often present in vital coalitions and have to be taken into account (Van de Wijdeven *et al.*, 2007). From the Rotterdam cases we know that several citizens, the front runners in the vital coalitions, were able to balance between these logics. They knew how to combine the logic of governmental procedures with the action radius of citizens. However, such successful 'entrepreneurs' are only one factor in good examples of vital coalitions. In urban areas the elements that make up a successful vital coalitions have been traced back to the following (Hendriks and Tops, 2002, 2005; Van de Wijdeven *et al.*, 2007):

- A sense of urgency
  There is a feeling that something has to be done (Hendriks and Tops, 2005). A certain tipping point has been reached (for instance the effect of a murder and firebombs in Hillesluis and bad publicity in Pendrecht). The circumstances do not necessarily have to be negative. Positive events can also create the energy to make change possible. The Christmas tree in Pendrecht was a common goal and gave residents the energy to take action for their neighbourhood.
- Entrepreneurship
  Recognising and acting on a sense of urgency requires a form of 'entrepreneurship'. 'Entrepreneurs' are often driven opportunists, people who are less inclined to comply with bureaucratic procedures but know how to gather support. They are able to communicate and mobilise, but can be good at looking for conflict as well if this supports their cause. Personality is therefore an important characteristic. Entrepreneurs tend to be prominent individuals with a capacity to act. They often know how to manoeuvre between the different logics.
- Government backing
  At key points in the process entrepreneurs need to be sure they have the backing of people that occupy powerful or influential positions within government, either political or administrative. This does not always have to be active. The presence or notion that people 'that matter' support the initiative ('shadow of hierarchy') is often sufficient (Scharpf, 1997). These officials, however, will have to ensure that the entrepreneurs have enough room for manoeuvre.

A vital coalition is a form of active citizenship and self-organisation, in which citizens and/or private or public actors take the initiative to act on behalf of a common concern or interest. Government takes a facilitating role and provides the necessary backing at crucial moments, but without taking over. The success factors for vital coalitions in urban cases are a sense of urgency, entrepreneurship and government backing.

3.5 Regional regimes and vital coalitions

Several similarities between regional regimes and vital coalitions have already been described. But there are also important differences. Regional regimes are much more extensive and established than vital coalitions. Vital coalitions are small-scale initiatives. They often start with a small and simple agenda, and not necessarily with the intention of becoming the regional priority. When comparing the components of an urban regime (agenda, coalition, resources and mode of alignment) to that of a vital coalition (see Table 3.1), it becomes clear that:
- the agenda of a regional regime has regional priority, whereas the agenda of a vital coalition need only be relevant to a few involved individuals;
- the regional regime has the ability to govern the regional area, whereas a vital coalition is (or at least starts out as) a small initiative;
- in general, a vital coalition can be traced back to certain individuals better than a regional regime, which is a larger and more complex interplay of actors;
- a regional regime has a certain amount of durability, whereas a vital coalition can be short-term too (for instance when the project is finished, the coalition may end as well).

Vital coalitions and regimes can coexist. In fact, it is the relation between the two that we are interested in. Whereas regimes represent the regional governing coalition and agenda, and the more or less vested regional interests, vital coalitions operate on a much smaller level, in the role of projects or niches. Niches are small initiatives for innovation, 'incubators' in fact.

Table 3.1. Vital coalition and regional regime compared.

	Vital coalition	Regional regime
Definition	A form of active citizenship and self-organisation in which citizens and/or private or public actors take the initiative to act on behalf of a common concern or interest	The informal arrangements by which autonomous and semiautonomous actors work together to make and carry out governing decisions relevant to a specific region (for instance urban/rural)
Agenda	Common interest or concern	Regional priority
Coalition	Citizens only or with private or governmental actors	Semiautonomous actors
Resources	Material and non-tangible	Material and non-tangible
Mode of alignment	Informal co-ordination	Informal co-ordination
Durability	Sometimes	Required

'Niches are locations where it is possible to deviate from the rules in the existing regime' (see Arts and Van Tatenhove, 2004: 345; Geels, 2004: 912).

If the rules and regulations of the regime indeed deviate from the vital coalition's aim, it is possible that the regime will hamper the development of the vital coalition. Sometimes the main reason for the establishment of a vital coalition is to react against the regime's current systemic structures. Certain individuals or actors may feel that the current governing context (the regional regime) does not adequately respond to their local or regional needs. However, discrediting the current regional regime is a very difficult process, since a regime in power has the advantage. Being in power, it has already proved that it can set the rules and dominate the agendas for a certain area.

Orr and Stoker address the issue of regime transition, a subject absent in Stone's work on Atlanta because in that case there was a relatively stable regime (Orr and Stoker, 1994). In Orr and Stoker's case of Detroit there is no stable regime, which leads the authors to develop a few points on regime change. They argue that new actors do not necessarily change the regime. Stone's Atlanta showed that regimes can cope with changes in personnel and do not have to change their way of working and the direction they are taking (Stone, 1989). Creating an alternative regime is problematic. To become dominant it requires a new set of material incentives and a new framework of meaning. Orr and Stoker develop a three-stage scheme for regime change. First, doubts are raised about the existing regime. This can happen when developments in the wider region appear to contradict or challenge the regime. These doubts then find a vehicle of expression in corporate, political or other potential leaders (note that this could be a vital coalition). Second, conflicts arise about a possible new direction for the old regime and the development of a new regime. This is a period characterised by much uncertainty and debate. Third, a new set of incentives and a new paradigm take shape, which marks the institutionalisation of a new regime (Stoker, 1995: 68).

It can be argued that regimes are protective and conservative because they do not want their power to be threatened or affected and will try to hold onto their structure and power. For a vital coalition to challenge the regime, it must therefore first have an appealing agenda that can attract the necessary resourceful actors. The regime may even provide the necessary resources for a vital coalition to function, in effect stimulating it. In Stone's Atlanta, subprojects that matched the regime's agenda were easier to develop than projects with other goals. Small businesses in particular were experiencing difficult times as they were outside the regime's coalition of business elite and black middle class.

Case studies are needed to determine the exact relationships between regional regimes and vital coalitions. Are regional regimes so dominant that they hamper the emergence of new initiatives in the form of vital coalitions, or are they the engine that makes innovative development possible in the first place? The relationship between regimes and vital coalitions is illustrated in Figure 3.2.

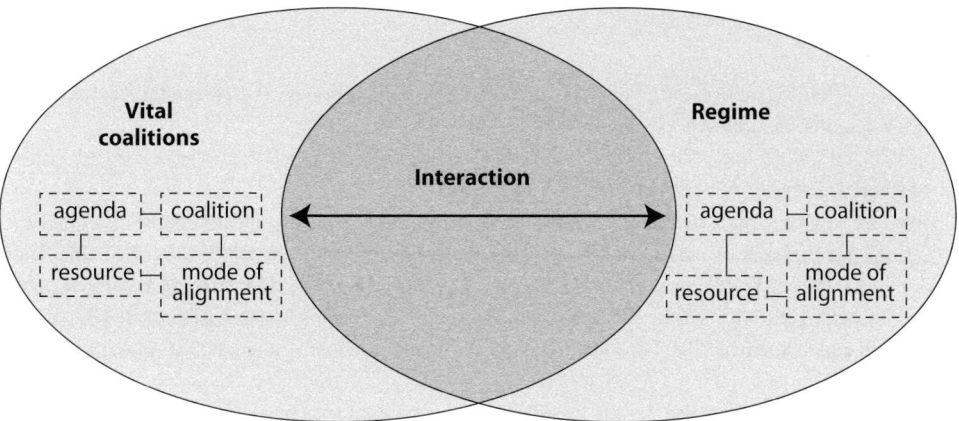

Figure 3.2. Interaction between the regime and vital coalitions.

3.6 Summary

In this final section we return to the central question of this chapter:

> *How can the concepts of urban regime and vital coalition be used to study public-private initiatives in a regional setting, how do these two concepts relate to each other, and what are conditions for vital coalitions?*

We have seen that regional regime and vital coalitions can be used to study how innovation in urban rural areas is promoted and obstructed. We have defined a regional regime as the informal arrangements by which autonomous and semiautonomous actors work together to make and carry out governing decisions relevant to a specific region (for instance an urban/rural region). A vital coalition is a form of active citizenship and self-organisation in which citizens and/or private or public actors take the initiative to act on behalf of a common concern or interest. Both can be seen as a form of policy network since both are built on the co-operation of autonomous and semi-autonomous actors around a common concern or agenda.

The regional regime has the advantage of having governing power, which a vital coalition does not have. If a vital coalition pursues an agenda similar to the regime's agenda, it may have less difficulty in developing it, and may even receive support. If a vital coalition has other goals, for instance because it feels that the current regime does not represent its interests well enough, it may clash with the regime and find its development obstructed, but if it is successful it can threaten the continuity of the current regime.

Case studies have shown that successful vital coalitions are always characterised by three factors: a sense of urgency, entrepreneurship and political or administrative backing. But whether vital coalitions in regional/rural context can achieve the same successes, and what that means for their relation to current governing structures (for instance, the regional regime), remains to be seen in empirical case studies.

References

Arts, A. and J. Van Tatenhove, 2004. Policy and power: A Conceptual Framework Between the 'Old' and 'New' Policy Idioms. Policy Sciences 37(3-4): 339-356.

Bang, H.P. and E. Sörensen, 1998. The Everyday Maker: A challenge to democratic governance (ECPR-paper). Aalborg University, Aalborg, Denmark.

Boogers, M., F. Hendriks, S. Kensen, P.W. Tops, R. Weterings and S. Zouridis, 2002a. Stadsbespiegelingen A: ervaringen en observaties uit het Stedennetwerk. Katholieke Universiteit Brabant, Tilburg, the Netherlands.

Boogers, M., F. Hendriks, S. Kensen, P.W. Tops, R. Weterings, R. and S. Zouridis, 2002b. Stadsbespiegelingen B: ervaringen en observaties uit het Stedennetwerk. Katholieke Universiteit Brabant, Tilburg, the Netherlands.

Bogason, P. and T. Toonen, 1998. Introduction: networks in public administration. Public Administration 76 (2): 205-227.

Börzel, T., 1998. Organizing Babylon – On the different conceptions of policy networks. Public Administration 76(2), 205-227.

Burns, P. and M. O'Thomas, 2004. Governors and the development regime in New Orleans, Urban Affairs Review 39(6): 791-812.

De Bruijn J.A. and A.B. Ringeling, 1993. Normatieve kanttekeningen bij het denken over netwerken. In: J.A. de Bruijn, W.J.M. Kickert and J.F.M. Koppenjan (eds.), Netwerkmanagement in het openbaar bestuur, over de mogelijkheden van overheidssturing in beleidsnetwerken. Rotterdam/Delft, the Netherlands, pp. 153-167.

De Graaf, L.J., 2007. Gedragen beleid: Een bestuurskundig onderzoek naar interactief beleid en draagvlak in de stad Utrecht. Eburon, Delft, the Netherlands.

Edelenbos, J. and R. Monnikhof, R. (eds.), 2001. Lokale interactieve beleidsvorming. Een vergelijkend onderzoek naar de consequenties van interactieve beleidsvorming voor het functioneren van de lokale democratie. Lemma BV, Utrecht, the Netherlands.

Elkin, S., 1987. City and Regime in the American Republic. University Press, Chicago, IL, USA.

Geels, F.W., 2004. From Sectoral Systems of Innovation to Socio-Technical Systems: Insight about Dynamics and Change from Sociology and Institutional Theory. Research Policy 33(6-7), 897-920.

Hamilton, D., 2002. Regimes and regional governance: the case of Chicago. Journal of Urban Affairs 24 (4): 403-423.

Hamilton, D., 2004. Developing regional regimes. Journal of Urban Affairs 26(4): 455-477.

Hendriks, F. and P.W. Tops, 2002. Het sloeg in als een bom: vitaal stadsbestuur en modern burgerschap in een Haagse stadsbuurt. Tilburg University, Tilburg, the Netherlands.

Hendriks, F. and P.W. Tops, 2005. Everyday Fixers as Local Heroes: A Case Study of Vital Interaction in Urban Governance. Local Government Studies 31(4): 475-490.

Horlings, I. and W. Haarmann, 2007. The soft stuff is the hard stuff; vital coalitions in rural-urban regions, paper for the ESRS conference, August 2007, Hungary.

Horlings, I. and W. Haarmann, 2008. Botsende belangen en vitale verbindingen, positionpaper over het integrated Project Nieuwe Markten en Vitale coalities Heuvelland. In: TransForum, Nieuwe Markten en vitale coalities Heuvelland Zuid-Limburg; innovatief praktijkproject, workingpaper no. 8, TransForum, Zoetermeer, the Netherlands, pp. 95-174.

Humphreys, D., 1996. Regime Theory and Non-Governmental Organizations. The Journal of Commonwealth & Comparative Politics 34(1): 90-94.

John, P., 2001. Local Governance in Western Europe. Sage Publications Limited, London, UK.

Junne, G., 1992. Beyond Regime Theory. Acta Politica 27(1): 9-28.

Kalk, E. and F. De Rynck, 2003. Burgerbetrokkenheid en bewonersparticipatie in de Vlaamse steden. In: F.

De Rynck (ed.), Voorstudies - Witboek Stedenbeleid: De eeuw van de stad. Over stadsrepublieken en rastersteden. Project Stedenbeleid, Ministerie van de Vlaamse Gemeenschap, pp.453-479.

Kickert, W.J.M., E.H. Klijn and J.F.M. Koppenjan (eds.), 1997. Managing Complex Networks. Sage Publications Limited, London, UK.

Lammers, C., 1983. Organisaties vergelijkenderwijs: ontwikkeling en relevantie van het sociologisch denken over organisaties. Spectrum, Utrecht, the Netherlands.

Lauria, M. (ed.), 1997. Reconstructing Urban Regime Theory: Regulating Urban Politics in a Global Economy. Sage Publications Limited, London, UK.

Leipzig Charter, 2007. Leipzig Charter on Sustainable European Cities. Prepared by the German Presidency of the European Union, May 2007 (www.eu2007.de, consulted on 14/1/2010).

Menahem, G., 1994. Urban Regimes and Neighbourhood Mobilization against Urban Redevelopment: The Case of an Arab-Jewish Neighborhood in Israel. Journal of Urban Affairs 16(1): 35-50.

Mossberger, K. and G. Stoker, 2001. The Evolution of Urban Regime Theory: The Challenge of Conceptualization. Urban Affairs Review 36(6): 810-835.

Mossberger, K., 2008. Urban Regime Analysis. In: J.S. Davies and D.L. Imbroscio, Theories of urban politics. Sage, London, UK, pp: 40-55.

Orr, M.E. and G. Stoker, 1994. Urban Regimes and Leadership in Detroit. Urban Affairs Quarterly 30(1): 48-73.

Peterson, P., 1981. City Limits. University of Chicago Press, Chicago, IL, USA.

Pierre, J., 2005. Comparative Urban Governance: Uncovering Complex Causalities. Urban Affairs Review 40(4): 446-462.

Reed, A., 1988. The black urban regime: Structural origins and constraints. In: M.P. Smith (ed.), Power, community and the city: Comparative urban and community research. Transaction, New Brunswick, NJ, USA, pp. 138-189.

Saartenoja, A., 2003. Can Urban Regime-Theory Help to Understand Urban-Rural Relations in the Regional Context, paper written for the Regional Studies Association in Pisa, Italy.

Sassen, S., 1991. The global city: New York, London, Tokyo. Princeton University Press, Princeton, NJ, USA.

Scharpf, F., 1997. Games Real Actors Play: Actor-Centered Institutionalism in Policy Research. Westview Press, Boulder, CO, USA.

Sellers, J.M., 2002. Governing from Below: Urban Regions and the Global Economy. Cambridge University Press, Cambridge, UK.

Sites, W., 1997. The Limits of Urban Regime Theory: New York City under Koch, Dinkins, and Giuliani. Urban Affairs Review 32(4): 536-557.

Stoker, G., 1995. Regime Theory and Urban Politics. In: D. Judge, G. Stoker and H. Wolman (eds.), Theories of Urban Politics. Sage Publications Limited, London, UK, pp. 54-71.

Stone, C.N., 1989. Regime Politics: Governing Atlanta 1946-1988. University Press, Lawrence, KS, USA.

Stone, C.N., 1993. Urban Regimes and the Capacity to Govern: A Political Economy Approach. Journal of Urban Affairs 15(1): 1-28.

Stone, C.N., 2002. Urban Regimes and Problems of Local Democracy. Paper prepared for Workshop 6, Institutional Innovations in Local Democracy, ECPR joint sessions, Turin, Italy.

Stone, C.N., 2004. The Governance Conundrum. Paper prepared for the European Consortium for Political Research. Joint sessions of Workshops No. 25, Policy Networks in Sub National Governance: Understanding Power Relations, Uppsala, Sweden.

Stone, C.N., 2005. Looking Back to Look Forward: Reflections on Urban Regime Analysis. Urban Affairs Review 40(3): 309-341.

Tops, P.W. and F. Hendriks, 2004. Governance as Vital Interaction: Dealing with Ambiguity in Interactive Decisionmaking. Paper presented at the International Conference on Democratic Network Governance, Copenhagen, Denmark.

Urbact, 2005. Citizen Participation Case studies. By the SecurCity Network, workshop 'Real Citizen Participation', Prague, Czech Republic.

Van der Stoep, H. and A. Van den Brink, 2006. Gespaard landschap 4. Ontmoeting met het Masterplan Stolwijkersluis. Tidinge van die Goude 24(2): 54-61.

Van de Wijdeven, T.M.F., E.M.H. Cornelissen, P.W. Tops and F. Hendriks, 2006. Een kwestie van doen? Vitale coalities rond leefbaarheid in steden. SEV, Rotterdam, the Netherlands.

Van Ostaaijen, J.J.C., 2010. Aversion and Accommodation. Political Change and Urban Regime Analysis in Dutch Local Government: Rotterdam 1998-2008. Dissertation research, Eburon, Delft, the Netherlands.

Chapter 4

Pathways for sustainable regional development: strategies and vital coalitions

Ina Horlings

4.1 Introduction

In Chapter 2 we described regional development in the Netherlands and the problems that occur. Innovative perspectives on rural development require new forms of co-operation between the state and civil society, but regional regimes can hamper sustainability if the search for new perspectives is trapped within traditional arrangements. The constellation of actors in regimes influences the way rural problems are defined. In other words, a regime is a 'value-inclusive system'. New actors have difficulty introducing their own problem definitions, participating in rural decision processes and co-operating with institutionalised governments. Their 'room for manoeuvre' is limited (Horlings *et al.*, 2009a). In situations gridlocked by the institutional context it is difficult to create capacity to act; the challenge is to develop new strategies and images that contribute to sustainability. The question is whether we can find new initiatives outside the formal context that can lead to vital coalitions at the regional scale.

What makes it more complicated is that the appropriate size for regional development can vary depending on geography, natural resources and amenities, skills and infrastructure (OECD, 2006: 114). The trend towards more decentralised policy throughout Europe is transferring responsibilities to subnational levels. At the same time, the greater attention devoted to place-based policies means there is more focus on the role of local entities in the implementation of such policies, and bottom-up approaches are being encouraged in several countries. This leads us to ask what obstacles there are to co-operation and co-ordination at the local level of rural policy, and what are the most effective mechanisms for enabling different local actors to work together. Partnerships in various European countries face a number of potential obstacles, such as the complex, rigid and fragmented national and supranational policies that affect rural development (OECD, 2006: 114, 127).

Faced with problems of effectiveness and legitimacy among a dynamic, highly educated and culturally diverse population, the challenge is to develop a new mode of planning. To anticipate and accommodate future opportunities more effectively, this should be more receptive to emerging actors and processes (WRR, 1989). A new perspective in spatial planning would entail a diversification and mobilisation of stakeholders and a cultural reinvention of the countryside for the development of a new rural economy (Mommaas and Janssen, 2008: 22). This transition process requires a reorientation of vertical and/or horizontal alliances or coalitions (both private-private, public-private and public-public) governing agricultural production and rural development. Our hypothesis is that specific *formal or informal networks* in the form of vital coalitions can contribute to innovation and sustainability in rural-urban regions. But can such coalitions be found in reality, and are they able to introduce new sustainable agendas? The central question in this chapter is:

> *Which types of coalitions can be identified in the context of Dutch regional development, to what extent do they introduce new sustainable issues and agendas on the regional scale, and which obstacles do they face?*

Vital coalitions have been defined in Chapter 3 as a form of active citizenship and self-organisation in which citizens and/or private or public actors take the initiative to act on behalf of a common concern or interest. In the context of rural-urban development these self-organising initiatives may be associations, interests groups, business communities or other groups who co-operate with other actors, such as government authorities, in order to set their agendas and achieve their goals. We concentrate here on strategies for sustainability followed by these bottom-up initiatives and the coalitions they form.

The questions that are addressed in this chapter are:
1. Which strategies for sustainable regional development can be identified in Dutch practice, illustrated by current regional projects?
2. What are stimulating and hindering regime factors in these projects?
3. Which types of coalitions can be identified and what are conditions for vitality?
4. What are the implications of the above for governance?

The aim is to offer insight into the organisational and institutional conditions for creating *capacity to act* in regional development processes. There is an urgent need for this, especially in regional government. Tops (2007) describes how governments are looking for a new relationship between collective and individual responsibility among their citizens because interactive policy-making processes tend to deteriorate into dead-end debates. These problems in regional governance cannot be explained from the perspective of networks alone, but require a regime approach and a means to create 'enabling power' (Tops, 2007). Citizens need to be given more responsibility and the opportunity to take it. What are needed are vital coalitions and new relationships between government and society (Tops and Hendriks, 2004).

4.2 Methodology

The empirical material is derived from information on eight Dutch cases. Private actors play an important part in these projects by creating new coalitions with public actors. The projects are initiated outside the formal policy context and can be seen as forms of self-organisation. The cases ensure a range of contextual variation in:
- their location in the Netherlands, in regions with different physical characteristics;
- the size of the project, varying from a new business to a regional planning process;
- co-operation between entrepreneurs, varying from a joint venture and a partnership to informal co-operation and coalitions with public actors;
- the sectors involved, such as agriculture, recreation and the medical sector;
- scenarios for sustainable development.

The data are derived from the literature and various documents about the projects, including several evaluation and monitoring reports, visits to the projects (TransForum, 2007a,b, 2008a) and interviews with private initiators of the projects, sometimes more than once. The interviews with respondents were structured loosely around topics like sustainability goals, networks, hindering and stimulating factors, and resources. For the analysis we used the urban regime theoretical framework, making a distinction between the topics agenda, coalition, modes of alignment and resources.

All projects, except those in the province of Noord-Brabant, were co-financed by TransForum, a foundation established by the Ministry of Agriculture, Nature and Food Quality and the Innovation Network to stimulate innovation in the agrosector and the countryside. TransForum stimulates projects in the Netherlands by co-financing the input of knowledge (TransForum, 2008b). One project was supported by Habiforum, which was established by public and private sector organisations to put multifunctional land use into practice and financed by five government departments.

The cases are listed in Table 4.1. They pursue new regional strategies, developing agendas that can be seen as expressions of different scenarios for the rural Dutch countryside, as described in Chapter 2:
- increasing the scale of agricultural production;
- an ongoing decoupling of agricultural production from the physical environment;
- multifunctional agriculture: diversifying agriculture with services such as nature and landscape management, water management and recreational and care services;
- rural-urban integration.

We examine these cases in Sections 4.3 to 4.6 and in Section 4.7 identify the types of coalitions formed in the projects, before assessing the contribution their agendas make to sustainability in Section 4.8. Section 4.9 summarises the factors that stimulate or hinder the initiatives. The implications of the diversified countryside and its consequences for governance at the European scale are examined in Section 4.10. Section 4.11 presents the main conclusions.

4.3 Innovative ways of increasing the scale of land-based agriculture

The scale of agricultural production has been increasing for fifty years or more in the Netherlands and is also a common strategy in other parts of Europe. New elements have been introduced into this strategy in recent decades, including:
- more extensive land use to reduce nitrogen losses per hectare, and regulation of mineral cycles;
- co-operation between agricultural and other agribusiness sectors, improving efficiency;
- vertical integration of businesses to gain more control of the food chain.

Increasing the scale of farm production can be a regional strategy and can be combined with sustainability goals in area restructuring projects. The case of the Overdiepse polder is an innovative example of this strategy.

Table 4.1. Regional projects in the Netherlands initiated by private actors.

Project and province	Scenario	Private actor	Strategy
Overdiepse Polder (Noord-Brabant)	Increasing scale of production	Overdiepse polder interest group	Regional design; responding to climate change by moving farms to higher ground
Sjalon (Flevoland)	Increasing scale of production	Sjalon business partnership	New organisational arrangements
The New Mixed Business (Limburg)	Decoupling of agricultural production from the environment (Industrial Ecology)	New Mixed Business partnership	Clustering of different agricultural sectors, integration and reuse of streams of resources
Northern Frisian Woods (Friesland)	Multifunctional agriculture	Northern Frisian Woods Association	Environmental management at regional level
Arkemheen Eemland Regional Innovation Centre (Utrecht)	Multifunctional agriculture/ rural-urban integration	Eemland Farm/Regional Innovation Centre	Upscaling innovation and knowledge arrangements
Agricultural care and quality products in Waterland (Noord-Holland)	Multifunctional agriculture/ rural-urban integration	The organisations Landzijde and My Farmer	Rural-urban contracts
Regional branding of Het Groene Woud (Noord-Brabant)	Rural-urban integration	Regional Co-operative Association	Regional branding
Regional development in Heuvelland (Limburg)	Rural-urban integration	Orbis Medical and Healthcare Group	New Markets approach

4.3.1 Regional design: farming on new *terpen* in the Overdiepse polder

The Overdiepse polder is an area of 550 hectares in the province of Noord-Brabant and has 94 inhabitants. The polder was selected as a search area for water retention to buffer the greater volumes of water and peak river discharges in the Bergsche Maas river, which are expected as a result of climate change. A sustainable solution to ensure flood safety is a combination of technical solutions (like dikes) and spatial measures. Spatial measures, as in the 'Space for the River' programme, aim to lower water levels upstream. The government's plans for water storage in this region focused on this strategy and created a clear 'sense of urgency'.

The Overdiepse polder interest group, a group of 16 farmers and one recreation entrepreneur, drew up their own plan to meet this requirement. Their starting point was that sustainable farming in this area meant that half the farmers can continue farming and the others will have to stop farming or move elsewhere. The plan is to build each of the farm buildings on a new *terp* (a traditional earth mound thrown up as refuges and for farmsteads in areas liable to flooding) and make provisions for flooding expected on average once in 25 years. Increasing the area of each farm will allow production to be 'extensified'. In the words of the initiator and chairman Sjaak Broekmans: 'Our plan is to create a sustainable situation. Because of the need for more extensive dairy farming, only half the farmers can stay. By limiting livestock density to two cows per hectare to keep manure production within the limits, you need 50 hectares for 100 cows.' The project contributes to sustainability in that it leads to more extensive land use and provides an answer to the rising water levels resulting from climate change. Extensive agriculture was not a project goal, but a means of adjusting farming practices to meet the requirements of the manure legislation. Besides guaranteeing continued farming in the area, the Profit aspects of sustainability in the plan are limited, and the People aspects are not addressed.

The group formed a coalition by working closely with the local and provincial authorities, both informally and formally. The project can be seen as an example of area development planning. Habiforum, a knowledge-based organisation, supported the initiative and helped to develop the plan (Van Rooy and Slootweg, 2003). The group was also supported by the farmers' union and won its annual entrepreneurship prize. The plan proved to be cost effective and well adapted to the policy goals, and thus created a 'window of opportunity'. Networking and informal contacts were crucial in the process. The group lobbied the government department responsible for water management and public works (*Rijkswaterstaat*) and government authorities at the local, regional and national level to get their plan approved. The private initiator, Sjaak Broekmans, maintained close informal contacts with the provincial executive councillor, which created opportunities for 'negotiations behind the scenes'. The executive councillor functioned as a policy entrepreneur.

The leader of the entrepreneurs participated in an official commission that supervised the project, providing direct access to local and farmers knowledge. Another benefit was that the group participated in the drafting of the environmental impact statement, and as a result local people did not oppose the final report. The group used its network power and gained influence by making effective use of the media and the project received much attention in the newspapers. The initiator also regularly delivered presentations about the project. Although the agendas of the governmental institutions and the coalition overlapped, some tensions with the current regime arose during the process (Van Rooy *et al.*, 2006). The environmental impact assessment procedure required the formulation of alternative plans, although all the participants had already agreed on the *terp* plan.

Another institutional obstacle was that government regulations still have to be complied with. In January 2007 the farmers had to double their manure storage capacity, despite the fact that the current farms will eventually be dismantled. Here the situational logic of the entrepreneurs clashed with the institutional logic of the government. There are also tensions between autonomous farm development and the long duration of the process. It turned out to be a 'mission impossible' to realise the plan within ten years, which was the original goal. Building work on the new farms will not start before the end of 2010, a big disappointment for the leading farmer. New government officials entered the arena, and some farmers drew up expansion plans for their farms, although all the farms will have to rebuilt when the plan is carried out, which will mean a loss of financial capital.

The inhabitants and entrepreneurs realise that they took an individual risk by co-operating in this project to help tackle a societal problem such as climate change. To avoid financial losses as a result of the project an arrangement was recently made to provide compensation for the value of the farmers' property. In 2007 a survey was carried out to find out which farmers wanted to stay and were prepared to invest. This revealed that one third wanted to stay, one third wanted to leave, and the other third had not made up their minds. Because the prospects for individual farmers differ, the initiator found it increasingly difficult to defend everyone's interests during the process and hold together a strong business community.

4.3.2 New organisational arrangements: the Sjalon New Large Farm

The Sjalon project is an illustration of the use of innovative arrangements for scaling up agricultural practices. *Sjalon* is Dutch for a surveyor's staff and the symbol of the group of agricultural entrepreneurs that started this initiative. Their assumption is that they can guarantee the continuity of their activities better by working together. Their motivations for starting the initiative were:
1. the falling prices of arable products;
2. the growing administrative burden and regulations on the environment, food safety and certification;
3. the limitations in company development;
4. new ideas about agricultural development, such as the need for innovation and new forms of co-operation.

The size of the farms in the Noordoostpolder, part of the new land in the province of Flevoland, is no longer suited to modern demands and the number of farms has declined. A group of entrepreneurs wants to amalgamate several arable farms to create a single 'New Large Farm' of 600 hectares, based on efficient use of labour and mechanisation. The aim of the Sjalon project is to establish a new form of co-operation between agricultural and non-agricultural entrepreneurs to enable more profitable primary production, more co-operation in the agricultural chain and delivery of societal services. The entrepreneurs will retain ownership of their own land, but will collectively manage all the land in the new farm. Co-operative arrangements with other sectors are not excluded.

The group wants to base their activities on People, Planet and Profit, although economic sustainability is the main aim. By 'Profit' they mean staying abreast of the latest technological developments and the investments these require. 'People' means focusing on good working conditions and the ability of arable farmers to derive pleasure and meaning from their work, and creating employment for young people. Under 'Planet' they aim to create a high-efficiency arable farming system with a low environmental impact. The project wants to contribute to societal goals. In the words of the initiator of the group Arnold van Woerkom: 'You have to take a leap forward to create a business that benefits the consumer, the landscape and society to gain a "license to produce" from the consumer'.

The initiator is the motor behind the group, a networker who is convincing and can inspire others, and when approaching potential members was sometimes even felt to be a bit threatening. He began with a brainstorming group, deliberately formed with entrepreneurs outside the sector, with different competences and views, to create added value. He created a 'niche', a space where people are able to think freely and creatively. The brainstorming group asked for advice from experts in the agricultural chain, finance and science. The brainstorming phase took a long time. Meeting every two weeks for four to five years, the group invested in building trust among themselves During this process two committed advisors pointed out their blind spots.

The group formed a foundation for the development of agricultural entrepreneurship (*Stichting ontwikkeling agrarisch ondernemen*) and set out to find a suitable legal structure for the new large farm. In 2006 they recruited seven members who were interested in the new concept. However, in 2007 four members left the group and progress stagnated and decisions were delayed. After one member who did not have the group's trust left, the discussion about the appropriate legal structure could proceed. The entrepreneurs hesitated for a long time, because to create one company they had to pool their machinery and labour, give up their independence and take financial risks. For potential new members, the concept was too concrete to allow them to introduce new ideas and too abstract to allow them to 'join a moving train'. It was not clear what the consequences were for farm succession, there was some distrust about the enthusiasm of the initiator, and emotions played a role, like the feeling that joining the new organisation meant stopping your own farm. Moreover, the legal details of the new company turned out to be very complicated. An obstacle was that some leaseholders were interested in the concept, but were anxious about losing their rights, and the public landowner, the State Property Department ((*Dienst der Domeinen*), part of the Ministry of Economic Affairs) was concerned about the fiscal consequences. This tiring process lasted from August 2004 to March 2008. Sjalon contacted senior civil servants at the Ministry of Economic Affairs and TransForum arranged meetings as an mediator. The final compromise was that Sjalon would obtain the leaseholders' rights; if in the future the initiative should stop, the farmers would get their rights back. Nevertheless, the farmers remained uneasy about this arrangement.

Sjalon was officially established as a trading partnership on 14 March 2008, with four members on three farms, still aiming to grow to 100 hectares. The advantages to the farmers of joining Sjalon are the differentiation of labour, allowing everyone to do what they can do best, the continuity of farming, fewer risks and greater stability. An external director will manage the business. Important network partners are a notary, accountants, the bank, government authorities, commercial partners and knowledge institutes (an institute for applied research and two universities). The group invested much in gaining the commitment of external organisations. The *Vrije Universiteit Amsterdam* (University of Amsterdam) drew up a business plan for Sjalon, with a subsidy from the provincial government, and the group looked for partners in the agricultural chain who could process their products.

There are plans to buy some more land as a joint business, but the price of land is high. The aim is to improve the multifunctionality of land use, quality and innovation. Some aspects still have to be developed, such as supply chain management, water retention, and a new farm management system, including new crops and crop rotations. The new company also wants to contribute to government plans on landscape management and the realisation of ecological corridors in this region. There are already regional plans for habitat restoration around the former island of Schokland near the project area and this region is also of cultural and historical interest.

Sjalon followed a selective strategy for finding actors, which saved energy. The Dutch regional agricultural entrepreneurs' association LTO Noord was critical about the initiative and was therefore not included in the process. A complicating factor was that the ex-chairman of this organisation was also an executive councillor at the local authority. The group wanted support from the Ministry of Agriculture and although the civil servants did not support the plan, the minister, a farmer himself, was positive. The ministry did not think it had a part to play, considering it to be a task for the provincial and local councils, but they put the initiative into contact with TransForum, which added environmental aspects and societal services to the agenda. The provincial authority is positive about the Sjalon plans.

Sjalon felt that the local authority had been somewhat obstructive and some council officers considered that the Sjalon plan was not innovative. The entrepreneurs complained that the local council was too passive and that their administrative capacity too small to handle as large a project as Sjalon. The most important obstacle was the differences between regional agendas. The local council wants to change the region's agricultural image and promote Nagele as a 'place for elderly people'. It also backs the diversification of the regional economy around the themes of housing, water and leisure. Telos research institute was asked to mediate and identify opportunities for development. Their conclusion was that the Sjalon plan 'can be integrated into a wider regional development context'. They concluded that the government authorities may be somewhat passive, but they are not obstructive. The resistance by government perceived by the entrepreneurs is more a product of the different perceptions and expectations of what governments and entrepreneurs should do (Hinssen and Hermans, 2009).

4.4 Decoupling and industrial ecology

The scenario of decoupling agricultural production from the physical environment is a globally inspired agrobusiness model. It is based mainly on the principles of industrial ecology and strives for efficiency and environmental gains. A regional element in the model is the proposal to use waste streams from arable production as inputs to intensive meat production. Reducing environmental damage by technological control mechanisms, closed systems and intensive land use are important features. Clustering several chain businesses in one location reduces energy use and transport costs.

4.4.1 The New Mixed Business: innovative clustering of different agricultural sectors

Efforts have been made to establish an Agropark in North Limburg since 2001. A highly innovative example of this model is the concept of the 'New Mixed Business' in the Agricultural Development Area in the municipality of Horst aan de Maas. This project is still in the planning stage. The proposed new complex will be extraordinary large for the Dutch context. As well as a co-digestion power plant and chicken slaughterhouse, the New Mixed Business will provide housing for 35,000 pigs and 1.2 million chickens. The concept is based on clustering different agrisectors to minimise transport distances, reduce fossil energy use and close energy, minerals and waste loops, with several production chains clustered at a single location to reduce transport of animals. The bioenergy plant will recycle manure and organic waste, producing energy and heat for use on the farm and for the sale of green energy to other businesses and to the national grid.

The participants in this New Mixed Business are developing new forms of co-operation as well as different types of sustainable technological innovation. The concept focuses on Profit (economic efficiency) and Planet aspects, such as reducing environmental problems and improving animal welfare (by less transport). The group gives five reasons for developing the concept:
1. the intensive livestock sector is fragmented;
2. the intensive livestock sector is vulnerable to animal diseases;
3. consumers' demands on societal goals are increasing;
4. international competition between farmers is growing;
5. there is a need for knowledge development and a knowledge infrastructure.

An advisory report (Kool *et al.*, 2008) has identified potential contributions to sustainability from a reduction in ammonia emissions, the on-site generation of energy, reduced emissions of greenhouse gases and, at the regional level, a significant reduction in odour nuisance. Veterinary risks will be lowered by the reduction in transport and by minimising the use of antibiotics. The project will generate employment and improve working conditions at the participating enterprises (see also Smeets, 2009).

Marcel Kuypers is a Dutch chicken farmer whose sustainability vision in agricultural production includes three main flows: (1) fossil energy, (2) the transport of artificial fertilisers, minerals and metals, and (3) vegetable (renewable) products such as cereals, maize, soya and sunflower. A shift is needed towards the third stream:

> *A sustainable agriculture makes an efficient use of chemicals and resources and provides intensive sectors with residual products. Waste minerals can be burned, fermented or gasified. An example: 18% of chicken food comes from imported soy scrap, this is a residual stream from the production of soya oil.*

Kuypers says the clustering requires a certain scale of production and a location within an agricultural development area. 'We want to make food products from chickens that meet consumers' needs by taking responsibility for a larger part of the production chain.'

The project was adopted by the Dutch Agrologistics Platform and by TransForum. This resulted in the establishment of a steering group, which included an innovation broker from the organisation Knowhouse, to support the project group. In 2004 the initiative was given a 'special status' and financial support from the agriculture ministry to stimulate the process. Despite this, the decision-making procedure has not gone smoothly. The initiative ran up against several restrictive spatial and environmental regulations and difficulties were encountered with obtaining the necessary permits. There was a discussion about the legal 'animal rights', which could not be granted because the new business was not yet operational and therefore did not have any animals. The entrepreneurs want to transport manure through pipelines, but this is not permitted under current regulations. Moreover, one of the technical aspects of the concept is to use the most modern technologies to clean emissions, but this new technique is not yet on the environment ministry's list of Best Available Technologies.

There were also difficulties with the spatial planning procedures. The zoning of areas for extensive, intensive and mixed farming under current planning policy in Noord-Brabant does not permit this project, which requires a large plot of eight hectares. A site therefore had to be found outside Noord-Brabant. The Agricultural Development Area in Limburg chosen as the new location for the project was initially designated to solve local problems, not to import animals from another province, which led to local resistance. In the words of the initiator: 'The question that came up was what is a farmer from Noord-Brabant doing in Limburg. These are gut feelings'. There was talk of locating the business on an industrial site, but this would require a major review of the spatial planning procedure and also sparked emotional objections, because intensive animal husbandry would then be regarded as an industrial form of production.

Much time and energy has been invested in informing and consulting the local population, especially after 2005, first by the entrepreneurs and later also by the local authority and the local Socialist Party. The chicken entrepreneur has the ability to inspire people and convince them of the benefits of the concept: 'The government asked me why I don't follow the beaten track? I said I am making a new one because it isn't there yet'. Several meetings with the local residents were held and scientists were invited by the local authority to give their opinion and analyse the political decision-making process (see Termeer *et al.*, 2009).

Despite all these efforts to inform the public, the project still faces local resistance and fears, as well as a general lack of acceptance of the concept. Intensive farming has a bad image in the Netherlands, fed by articles in the press and citizens protests. Between 2005 and 2009 the controversy surrounding the project intensified. The local Socialist Party formed a coalition with citizens who did not want this agricultural complex 'in their back yard', and in 2007 local inhabitants protested against new 'megafarms' and formed the *Behoud de Parel* ('Conserve the Pearl') action group. The media have devoted much attention to the New Mixed Farm, especially the local media, and their use of terms like 'megasheds' and 'pig flats' have had a considerable influence on the image of the project and the decision-making process (Termeer, 2009: 28).

The technical sustainability concept clashed with the values of citizens concerned about landscape quality and animal welfare. In response, the council carried out a sustainability study (Kool *et al.*, 2008). On 8 July 2008 the local authority approved the New Mixed Farm's application for a location in Grubbenvorst on condition that they meet some additional sustainability criteria. These criteria were the result of political pressure, but the entrepreneurs considered these additional constraints to be unfair. To remove uncertainties and give due justice to the objections that have been raised, the council launched a considerable additional research effort, including the sustainability study, a health effect screening, a landscape plan and a financial evaluation of the business plan. The local authority also wants to make a formal and binding contract with the entrepreneurs to ensure the additional criteria will be met. The result is that entrepreneurs are having to make considerable investments in money and time to win over the politicians, environmental organisations and citizens.

The discussion was scaled up to the national level when national environmental organisations became involved. The Telos research institute was asked to analyse the arguments and to identify the remaining room for manoeuvre for this agropark (Horlings, unpublished data). At the beginning of 2010 the outcome was still uncertain.

4.5 Multifunctional agriculture

In this vision, agriculture focuses on multifunctional sustainable land use. Food production is just one of the functions of agriculture, along with 'green and blue' services like nature and landscape management, agrotourism, day care for special groups of people, and water storage.

4.5.1 Regional environmental management by the Northern Frisian Woods (NFW)

An example of this strategy is the Northern Frisian Woods Association, a group of farmers who have adapted their farm management to reduce mineral inputs and benefit nature and the landscape. They also manage the *Noardlike Fryske Wâlden* (Northern Frisian Woods) National Landscape with other organisations and the provincial authority. The Northern Frisian Woods (NFW) is a combination of four farmers' associations that see themselves as initiators and producers of 'farm-oriented regional development'. In an area of approximately 55,000 ha about 750 farmers and the provincial authority are jointly developing an agricultural system that will safeguard the landscape and ecological features and generate an acceptable income for farmers. The goal of the NFW project is to combine landscape, nature and environment in a 'regionally characteristic way' to deliver long-term, socially and economically sustainable agricultural management.

The key is to achieve natural and environmental objectives through a collective approach at the regional level. This can be done technically, for example by using different fertiliser distribution methods, but may also be approached legally, by acting as a single legal entity for which just one environmental permit is sufficient. The group wants to develop a regional contract as a framework for regional steering, with integrated regional co-operation on landscape and nature management, creating opportunities to increase biodiversity and the quality of the recreational offer. The farmers are trying to lower their mineral losses at the regional level and maintain the small-scale landscape in 85% of the area. The group contributes mainly to the Planet aspects of sustainability, by making reductions in mineral use and carrying out habitat management and landscape maintenance work, and less to the Profit and People aspects.

It is an innovative experiment in self-regulation, with goals monitored at the regional level rather than at individual businesses. The project is supported by TransForum. The farmers have been working closely with the Department of Rural Sociology at Wageningen University (WUR) for about 20 years, and have formed a coalition with governments at different levels: local councils, the Friesland Provincial Council, two government departments and several

NGOs. Government authorities play a supporting role, especially the provincial government. The NFW is part of a steering group that functions as the National Landscape management authority, and is chaired by a provincial deputy. The direct line of influence from the WUR to the former agriculture minister created temporary 'room for manoeuvre' for farmers by allowing them to spread manure onto the soil in exchange for lower inputs, but national and European manure and fertiliser regulations remain an important obstacle. The group furthers its aims by effective networking and use of the media. The wilfulness of the farmers in fighting for their own region, is a stimulating cultural factor. The NFW's ambition of playing a leading role in regional management led to friction with environmental organisations and local councils who wanted to remain in control, while shifts in responsibility from public to private actors tends to generate resistance. According to Douwe Hoogland, chairman of the Northern Frisian Woods Association, the challenge for the near future is to develop the region economically and to broaden the 'farmers' steering' to 'regional steering' in co-operation with governments and NGOs, without losing the initiative or becoming institutionalised:

> *We are so closely linked with the local councils now that we have to interact in a good way. An important challenge is also to strengthen the Profit aspects of sustainability by generating more income in the region, by working together with non-agricultural small and medium-size businesses.*

4.5.2 Upscaling innovation and knowledge arrangements: Regional Innovation Centre

The Eemland Farm (*Eemlandhoeve*) at Arkemheen Eemland, in the province of Utrecht in the middle of the Netherlands, is an educational and multifunctional farm. It can be considered as a diversified form of agriculture, combined with new alliances between town and country. The focus of the Eemland Farm is on Planet (sustainable land use) and Profit (professionalisation of multifunctional agriculture). A multifunctional farm in itself is not innovative, but the company is also a motor for developing new rural concepts and coalitions on different spatial scales. The entrepreneur Jan Huijgen, who was also the initiator and chairman of the farmland conservation association in this region, started a national 'rural-urban co-operation' with other nature conservation associations to stimulate multifunctional agriculture in the Netherlands. Their goal is to share ideas and resources so that new initiatives do not have to start from scratch.

The Eemland Farm was one of the initiators of the TransForum project Green Valley, which aimed to link agricultural entrepreneurs in three provinces and support their transition to sustainable multifunctional agriculture. The definition of sustainability in the Green Valley project includes not only People, Planet and Profit, but also Partnership (new coalitions), Profession (aimed at diversifying agriculture), Place (developing a regional market) and Principles (values and sense-making). The project faced difficulties in creating a firm business community and it turned out to be very difficult to align three provincial authorities around this project. The Green Valley project also revealed the risk of introducing innovative ideas, of getting too far ahead of the pack, and it had to compete with other organisations in the region that also wanted their share of influence (Horlings and Van Mansfeld, 2006).

The Eemland Farm was reconstituted as a Regional Innovation Centre on 5 March 2008. Its purpose is to collate expertise to answer questions that arise in practice, and to stimulate innovation and integration with student education by establishing a 'Rural Academy' (see Roep *et al.*, 2009). The entrepreneur Jan Huigen was the driving force behind a conference at the Eemland Farm on 'The city in search of farmers', where the agriculture minister, Gerda Verburg, was presented with the Amersfoort Agreement on implementing concrete projects to strengthen the relation between the built-up area and surrounding green areas. The aims were scaled up by organising a European conference on regional development in October 2008, bringing different national and European rural and knowledge institutes together around the theme 'E-motion for a versatile countryside'.

4.6 Rural-urban integration

In rural-urban regions we can see a scaling up of functions, spurred on by global forces, which makes landscapes more monotonous and erodes the physical differences between regions. The influence of urbanisation and the intensification and the upscaling of agriculture makes it more difficult for regions to remain distinctive.

Parallel to the process of globalisation and the decoupling of agriculture from ecological conditions, a process of re-regionalisation is underway (Horlings, 1996), strengthened by the growing urban anxiety about the degradation of landscapes. The scarcity of valuable landscapes in areas where they can satisfy the urban demand for rural values like space and peace and quiet is feeding a growing interest in regional qualities (Horlings *et al.*, 2006b). Consumers live, work and recreate in expanding networks, but still feel committed to their own region. The rural domain is no longer exclusively for food production, but in some regions is being transformed into a 'consumption landscape' to meet people's demands. This is part of a broader trend referred to as the 'experience economy' (Pine and Gillmore, 1999). Rural experiences embodied by regional quality products, walking in the countryside, attending country fairs and regional festivals, and an interest in dialects and cultural history are all expressions of this trend. The landscape is increasingly becoming the 'experience space' of urban society.

4.6.1 Rural-urban contracts in Waterland

This urban-inspired vision starts from the view that the boundaries between urban and rural functions are eroding. Sites for healthcare facilities and housing are being sought in the countryside, while farmers co-operate with urban partners to stimulate sales of their regional products. In this sense a sustainable strategy means the creative use of urban economic driving forces (Profit) to maintain the quality of the cultural landscape and current land uses (Planet).

An example is the Waterland region near Amsterdam, where the *Landzijde* ('Landside') organisation launched the 'Green Care Amsterdam' project, which seeks to realign city and countryside. Agriculture in Waterland, a fen meadow area of high landscape quality, is under pressure and is looking for new sources of income to benefit from the proximity of Amsterdam. The local farmland conservation association, 'Water, Land and Dikes' (*Water, Land en Dijken*), is active in maintaining the landscape. It is one of the oldest agricultural farmland conservation associations in the Netherlands and was established in the 1980s (Horlings, 1996). Landzijde is a commercial business and knowledge centre on farming and care that co-ordinates the placing of a variety of clients needing day care at 87 farms.

The Landzijde project 'Green Care Amsterdam' started in 2008. The farmers in Landzijde work with the City of Amsterdam to provide care, shelter and day activities, especially for people with psychological problems and reintegrating drug addicts. The project also makes farms suitable for special education needs through a learning programme. With help of TransForum, an alliance was established with researchers on 'care farming' at the University of Wageningen and the department of psychiatric research at Radboud University Nijmegen and the University of Amsterdam, to monitor the health effects. From a 'learning history' of Landzijde by the University of Amsterdam we can conclude that it formed a network that was able to align different worlds, agriculture and the care sector. The spin-off was the acquisition of clients and care organisations, which offered major advantages for farmers (e.g. training opportunities and less bureaucracy). Landzijde promotes professional 'care farming' and works to guarantee the quality of care, but also tries to maintain the identity of farming and the small scale of the services offered on the farms.

The sustainability gains of the projects are mainly on Profit aspects, by creating new income sources for farmers (Landzijde is a commercial organisation), and People aspects, by caring for different clients in rural as well as urban areas. The initiator Jaap Hoek Spaans, director of Landzijde, is an optimist and has not encountered any difficult institutional barriers so far. Although there are some differences of opinion between actors, and in the beginning the farmers' union was very doubtful about the project, intensive lobbying of the local councils has generated support. There is a strong personal connection between the farmland conservation association, Landzijde and the regional organisation Green Hat, which markets and sells regional food products for 100 farmers. Green Hat and Landzijde are now developing new urban-rural alliances. One of these is the food project 'My Farmer', which resulted in the opening of a co-operative supermarket in the centre of Amsterdam in 2008, selling regional products from Waterland and other quality products. The left-wing Amsterdam executive councillor supports the project because it fits in well with Amsterdam's city promotion and its recent food strategy, inspired by London's food strategy.

The close links between the projects are a stimulating factor in the region, but lead to dependency on the work of the main initiator Jaap Hoek Spaans. Now the pioneering phase is over, he is looking for a new Landzijde director. The initiator has a large network and played an important role as chairman of the national 'Versatile Countryside' task force supported by the Ministry of Agriculture, Nature and Food Quality, which stimulates multifunctional agriculture.

4.6.2 Regional branding in Het Groene Woud

The competition between regions is growing and they are becoming increasingly dependent on the image they communicate to make themselves more distinctive. One of the strategies regions pursue is regional branding, following the trend of city branding. Branding is the development and marketing of regions around their core values, both in a physical sense (quality of the landscape) and in an economic, social and cultural sense. Social bonds, historic buildings, cultural heritage, language and the stories told are all part of regional identities. Branding can be seen in the wider context of the dependence of regions on their 'cultural load', of the sense of belonging they are able to create, not only among visitors, but also entrepreneurs, investors and inhabitants. Sustainable development in this sense means marketing the region and developing new product-market combinations (Profit), based on the quality of landscape and biodiversity (Planet) and co-operation between the actors involved (People).

Based on a comparison of six European regions, Árnason *et al.* (2009) states that the process of branding in networks can be divided into three aspects: (1) increase visibility, (2) develop new products, and (3) reorganise activities. We can see these aspects in the case study of *Het Groene Woud* ('The Green Forest') National landscape, which lies between the cities of 's Hertogenbosch, Tilburg and Eindhoven in the south of the Netherlands. The landscape contains a network of nature conservation areas, valuable landscapes and several villages. Historically Het Groene Woud is a mix of different landscapes, including the Meierij, Campina, Mortelen and the Kempen. The label Groene Woud was introduced by nature conservation organisations who wanted to improve the quality and connectivity of natural habitats in this region. It is a rural-urban area with 1.5 million inhabitants and a concentration of top quality knowledge-intensive businesses (industry and business services). Urban functions are becoming more important, while agriculture has to adapt to urban demands, such as maintaining an attractive recreational landscape. The challenge is to strengthen the regional economy while preserving landscape quality.

Sustainability is this sense is combining Planet (mainly landscape quality) with prospects for the entrepreneurs and employment opportunities. There is attention for the People aspect in the sense that co-operation and voluntary work is valued high in this region. The initiator of the branding process Frans van Beerendonk, is a strawberry producer. As a local inhabitant, entrepreneur and former board member of the Federation of Agriculture and Horticultural Organisations (LTO Nederland), he wanted to ensure that the many spatial plans for this area still leave room for entrepreneurship. In his opinion, economic activities, when developed in a coherent way in networks and chains, are the foundation of regional development and contribute to regional identity and sustainability. He introduced the idea of regional branding to *Innovatieplatform Duurzame Meierij* (IDM), a regional organisation for innovation which also co-ordinated EU Leader+ projects. In July 2003, some members of IDM took part in a trip organised by the Noord-Brabant provincial authority to Cork in Ireland to learn from the Fuchsia initiative in this region. There they came into contact with the chairman of the West Cork Leader Co-operative Society, which has more than ten years experience with regional branding to strengthen the regional economy of southeast Ireland.

This coincidental meeting sparked the idea of an international exchange visit, which took place in July 2003 in Ireland. More than forty people from the private and public sectors in Noord-Brabant took part, as well as a group from the South Downs in the UK who are marketing regional products. IDM decided to go 'with a bus full of people' to mobilise broad support for this idea and created a basis for co-operation. Some regional meetings were also organised to attract entrepreneurs. Some people from the group also visited other regions in the Netherlands, including the province of Drenthe, to learn about their ideas on marketing regional products.

Events were organised in the region to raise awareness among the population. In 2006, 2007 and 2008 a large regional festival was organised to promote the attractions of the region and activities in Het Groene Woud, which were co-ordinated by the Regional Festival Foundation, chaired by the strawberry farmer. After the trip to Cork a group of entrepreneurs was formed to turn the concept of branding into commercial opportunities and market regional products. A label for their products was launched in November 2007 and in 2008 a commercial co-operative association with 30 members was formed to market the regional products, and a further 160 interested entrepreneurs were asked to join the co-operative. With help of a commercial advisor, who also feels personally committed to the region, the group developed a business plan to develop broad sustainability criteria at the business as well as the product level. Their first step was to make agreements with farmers about habitat management and development. In 2010 50 farms were certified and now produce under the Het Groene Woud label. The group then sought co-operation with other regions developing similar activities, such as the Kempen region and the Maashorst in Noord-Brabant. The project in Het Groene Woud also wants to develop new relations between the region and the triangle of cities around it: Den Bosch, Eindhoven and Tilburg.

A national innovation in this region was the establishment of a 'regional financial account', initiated by an executive councillor of Boxtel, Ger van Oetelaar. It works like a regular savings account. Public and private actors can open an account (Isis account) and receive the normal interest. The bank adds a further 0.15% to a regional fund, the Horus Fund, so 'people can give money without giving money'. Account holders can also donate interest to the fund, gifts or specific legacies. The aim is to spend 200,000 euros a year on projects in Het Groene Woud. The Fund has been successful and can be seen as one of the most innovative ways to finance landscape quality. The regional account has a credit of 40 million euros and the stream of capital is used to fund a subsidy programme to stimulate sustainable projects in the region. So far 400,000-500,000 euros has been invested in projects (Beckers *et al.*, 2009).

An obstacle in the initial years was the lack of government leadership in co-ordinating the huge number of plans, initiatives, ideas and projects. The bottom-up process created energy and generated many initiatives, but also led to fragmentation and a lack of coherence. Initially the specially formed regional steering platform of local councils did not show much ability to carry projects through. Recently, however, in response to questions that arose in the region, the provincial authority, as co-ordinator of the new National Landscape, took the lead in restructuring three organisations to form one new regional council: IDM, the regional steering platform, the reconstruction committee and the organisation for part of the area, the Loonse en Drunense dunes. Despite this, the urban-rural governmental divide persists: the cities are working together in 'Brabantcity' and their own initiatives and their plans are poorly co-ordinated with plans for the National Landscape, and the nature conservation organisations and entrepreneurs ('the green and red track') are not yet firmly allied around a joint agenda. The structure of the provincial organisation is divided into different programmes, which sometimes makes it difficult to obtain financial resources to keep the process going.

4.6.3 The New Markets approach in Heuvelland

The quality of the landscape in the 'metropolitan areas' is under pressure. Urban development has a growing influence on these regions. Instead of being outside the city, rural areas now find themselves 'in between' urbanised areas. The expansion of urban activities and the housing market are driving a suburbanisation of work and supply functions as urban-rural boundaries disappear (Verwijnen and Lehtovuori, 1999).

How can the environmental quality and values of landscapes be maintained? The New Markets strategy tries to link economic developments to spatial networks and to attract large rural and urban entrepreneurs to invest in the landscape. The aim is to develop new markets based on alliances between different sectors, such as food, recreation and care, combining economic investments (Profit) with landscape quality (Planet).

Heuvelland, which means 'land of hills', is a metropolitan landscape in the south of the Netherlands between the cities of Maastricht, Heerlen and Sittard. Heuvelland is a National Landscape and one of the oldest tourist regions in the Netherlands. The landscape is characterised by rolling hills, forests, brooks, sunken lanes and meandering rivers. Both agriculture and tourism are putting increasing pressure on this landscape, but international market forces are squeezing profits: the ongoing industrialisation of food production is lowering the costs per unit, and low-cost airlines allow people to take cheaper holidays outside the Netherlands. It was feared that, if left to their own devices, both tourism and agriculture would gradually become marginalised and no longer be able to act as co-producers of the landscape (ZKA *et al.*, 2005; Mommaas and Janssen, 2008). Heuvelland is becoming a 'post-productive landscape', where new functions related to urban needs are becoming more important as drivers of the economical vitality of the region.

The *Heerlijkheid Heuvelland* project was started in 2003 to improve the recreation economy of the region. The approach called 'New Markets' tries to develop new product-market combinations that function as new economic drivers of landscape quality. In the first phase the lead role was taken by the non-governmental regional development organisation (LIOF), with advice and knowledge support from various partners, including Urban Unlimited, ZKA Advisors and Tilburg University. The regional government supported the process, but did not actively participate in the first phase. The outcome of the desk research was translated into 'opportunity maps', a new and innovative method to identify and visualise new opportunities for development by combining potential markets with spatial networks (ZKA *et al.*, 2005) The maps were presented to entrepreneurs during round table meetings organised by LIOF to identify new potential markets. In Heuvelland only a selected group of stakeholders, mostly large entrepreneurs who are potentially committed to invest in the region, were invited to take part. The outcome of this phase was the identification of five development opportunities. Groups of interested entrepreneurs were assembled to develop each of these opportunities. One of these was 'Healing Hills', a coalition between Orbis, a regional medical and care concern, and the hospitality company Château Hotels, who signed a contract with a national insurance company for two holiday and healthcare packages allowing surgery patients to recover in a 3 or 5 star hotel set in an attractive landscape.

The Heuvelland project as a whole was actor oriented rather than plan oriented, and concentrated on entrepreneurs who wanted to think and act strategically. The group of selected entrepreneurs was kept small to reduce complexity in the field of action; only those who had a stake in the problem and were willing to help to solve it were recruited (Mommaas and Janssen, 2008: 31). This 'selective mobilisation' approach is rather new in the Netherlands, where broad, consensus-seeking negotiations between interest groups are part of a long Dutch tradition.

From an evaluation of the process some conclusions can be drawn about the stimulating and hindering aspects (Horlings and Haarmann, 2007, 2008; Van Mansfeld and Van der Stoep, 2007). In regional processes in metropolitan landscapes, the Profit and Planet aspects of sustainability are often at odds in the search for ways to link new economic perspectives to spatial quality in an urbanising environment. This region was no exception. The New Markets approach may have come too soon, as a joint agenda on the future of the landscape had not yet emerged. There is still a risk that the project will follow a one-sided economic agenda, focusing just on new economic markets in the short term.

Although a public-private coalition was formed, the dominant institutional setting remained intact and existing interests were not discussed. Another hindering factor was the divide that still existed between urban and rural policy and between vertically organised policy domains. Recreational entrepreneurs saw the importance of landscape quality in attracting tourists, but the combination of private and public goals was not obvious to them. It is too early to say whether the New Markets approach will lead to new investments in the landscape, but the approach has encouraged a new mindset. Private co-operation turned out to be achievable at the project level where entrepreneurs can realise their private interests, but scaling up to the level of regional development, where the agendas and interests of a great variety of actors clash, proved difficult. In regional processes like this, tensions arise between different types of logic (Horlings *et al.*, 2006a). In this case the 'institutional logic' of the provincial government was that entrepreneurs should take on more public responsibility and establish an organised business community. The 'situational logic' of the entrepreneurs was their desire to implement concrete projects, with financial and institutional support from the government. This project also showed that that 'the soft stuff is the hard stuff': not only facts, but also feelings, emotions and personal relations play a crucial role. The conditions needed for organising regional coalitions have been shown to be stimulating trust, creating networks and explaining the different interests, images and expectations from the beginning (Horlings and Haarmann, 2007b).

The follow-up to Healing Hills is the development of integrated care community (ICC) concepts: care facilities in a village or town or in the countryside that are integrated with other regional services like recreation, wellness, childcare and food supply, and with the landscape. The initial idea for a programme for 'healthy environments' was proposed by the regional office of the medical company Orbis at a meeting of the stakeholders of the New Markets project during the first phase in 2006. In 2008, when New Markets was having difficulty in generating synergy between economic development and landscape quality, concepts were developed for ICC in the urban and rural context. However, when the financial crisis broke, Orbis had to pull out (Haarmann *et al.*, 2009). The Heuvelland case study is analysed more in depth in Chapter 6.

4.7 Types of coalitions

Some of the projects we have described illustrate the tension between the still dominant traditional agricultural regime and new agendas. While the concrete form in which these tensions become manifest is shaped by regional characteristics, the cases show that there are some common elements in the coalitions formed:
- The actors involved introduce new strategies, such as regional branding, regional farm management, new organisational arrangements and forming alliances with non-agricultural sectors.
- Public-private co-operation on different levels is established with local and provincial authorities, and sometimes national governments.
- The projects pursue an innovation agenda by forming coalitions with knowledge institutes, which function as intermediary organisations between science and practice. This is not surprising, because most of the projects were already linked to TransForum or Habiforum, organisations that support empirical innovation with knowledge.
- Alignment of actors is often improved by informal contacts and 'negotiations behind the scenes'.
- The actors use resources like time, money, knowledge and power. The coalitions use discursive and organisational power to reach their goals. Private actors strengthen their power bases by forming a 'business community', networking intensively and using the media.

The cases illustrate different forms of coalitions and a variety of mechanisms. Three types of coalitions can be identified, although these overlap in practice (Table 4.2). Which form of coalition is appropriate depends on the problem definition and the regional situation. We do not yet know which type of coalition will be most effective in the long run because most were formed only in the last few years.

To what extent can the coalitions described be defined as *vital* coalitions? By vitality we mean energy and productivity to create capacity to act in order to change regional agendas, realise goals or change the formal or informal 'rules of the game'. The projects are vital in the sense that they influence local or regional agendas, for example in Waterland, Northern Frisian Woods, New Mixed Business and Het Groene Woud, and are to a certain extent able to create room for manoeuvre to realise (at least partly) their goals. In most cases they participate without adapting the dominant regime in their region, but pursue their own innovative path. These projects suggest there are three crucial conditions for vital coalitions. First, a strong *business community* (co-operation between private companies) with a joint agenda, in which private initiators with passion work as 'leaders of change' (see also Chapter 5). Second, a *sense of urgency* in the region, creating synergy between private initiatives and public goals and plans, combined with good timing to create a 'policy window'. Third, working in *formal/informal multilevel networks*, negotiating 'behind the scenes' and making use of power and the knowledge of research institutes. Changing the rules of the game has so far been 'a bridge too far' and the current institutional context has been *challenged but not changed*. For example, the Northern Frisian Woods has been questioning national and European manure and fertiliser regulations since the beginning of the 1990s and the New Mixed Farm is still having difficulty in negotiating the current regulations.

Table 4.2. Types of coalitions.

Type of coalition	Characteristics	Examples	Goal
Business-oriented coalitions	Coalition between different agricultural or non-agricultural sectors	New Mixed Business, Sjalon, New Markets approach	Create win-win situations based on joint interests in economic development
Rural-urban alignment	Coalition between entrepreneurs and consumers/inhabitants of cities	Branding Het Groene Woud, Regional Innovation Centre, Green Care in Waterland	Create new linkages and commitment between producers in rural areas and citizens by producing products and services
Steering coalitions	Coalition of entrepreneurs, government authorities and NGOs at the regional level	Overdiepse Polder, Northern Frisian Woods	Negotiate and influence government decision making by participating in formal and informal organisations, ad-hoc project groups or more permanent steering groups

4.8 Contribution to sustainable development

How do the agendas of the initiatives contribute to sustainable regional development? The coalitions mostly define sustainability in a broad sense, including People, Planet and Profit aspects. However, in practice most projects focus on economic aspects and the People aspect is generally underdeveloped. The claim of sustainability helps in getting support from government authorities and financial support from organisations like TransForum. In some projects sustainability seems to be little more than 'window dressing' to legitimise the project, rather than being an integral framework for development.

Having said this, we see that the agendas of the coalitions include innovative elements that contribute to sustainable development. The concept of *regional branding* as perceived in Het Groene Woud tries to market the region from an integral perspective. The goal is not merely to develop quality criteria for products or develop a regional label, but also to stimulate co-operation between 'red' and 'green' sectors, and to rebalance regional costs and benefits using the concept of the 'regional account'. The regional account itself is a novelty, initiated by a 'public entrepreneur'. The producers are expected to invest in sustainability at the product or business level and, if they can, contribute financially to the regional account. The New Markets approach tries to form *new innovative alliances between agricultural and non-agricultural sectors* that contribute to landscape quality. The approach links economic changes to the core physical qualities of the region in 'opportunity maps'. In the Overdiepse polder a creative *regional design* was developed to try to combine functions that are usually spatially separated, such as agriculture and water storage. The *clustering of different agricultural activities*, as developed by the New Mixed Business, can be seen as a form of industrial ecology, based on assumptions of ecological modernisation. This perspective has potential sustainability credentials, although rather narrow and technologically defined. The sociocultural aspects of sustainability are underdeveloped, which led to protests by citizens. Sjalon, Landzijde and Northern Frisian Woods developed innovative forms of *private organisation*, such as new forms of business communities and collective arrangements for more sustainable agricultural activities and farm continuity. In the case of the Northern Frisian Woods, the responsibility for mineral management changed from an individual to a collective, regional responsibility.

It is too early to say what the concrete and measurable effects of the projects will be in terms of sustainability. The only project that is being systematically monitored for sustainability is the Northern Frisian Woods (see Van der Ploeg *et al.*, 2003). Their approach has led to a robust reduction of nitrogen losses (Renting and Van der Ploeg. 2001), and 30 farmers are trying to go further and reduce their ammonia emissions to below the legal standard (TransForum, 2007b), the soil biology has improved (De Goede *et al.*, 2003) and 850 farmers, farming 85% of the area, participate in nature and landscape management schemes. It would be instructive to have a system for monitoring the sustainability of all the projects at the regional level.

Projects like Sjalon, Het Groene Woud and New Markets show that it is difficult to scale up projects to the regional level and to attract investments from entrepreneurs in public goals, such as landscape quality based on new market opportunities. The New Market approach is based on the assumption that non-agricultural economic sectors can function as new carriers for landscape quality. However, in regional processes in metropolitan landscapes such as these, tensions often arise between Profit and Planet aspects of sustainability in the search for ways to link new economical perspectives to spatial quality in a context of urbanisation. There is still a risk that the project will follow a one-sided economic agenda, focusing just on new economic markets in the short term (Horlings and Haarmann, 2007b).

4.9 Stimulating and hindering factors in regional development

We have reviewed the stimulating and hindering aspects in the regional development projects using the framework used in regime theory, which we introduced in Chapter 2. We analysed the following aspects (see Table 4.3):
- agenda
- coalitions (actors, roles and co-operation);
- modes of alignment (contacts, rules, personal aspects);
- resources (skills, time, knowledge, money, power).

Some of the obstacles can be traced back to the working of regimes, essentially the institutional and policy context. Hindering factors were found to be:
1. *Organisational policy constraints*. In some projects, such as Heuvelland, a fragmented provincial government organisation with different departments for 'red' and 'green' policy obstructs co-operation. Another obstacle is the difference between the situational logic of entrepreneurs and the institutional logic of governments. Public and private action work to different time horizons. Entrepreneurs often find the rate of implementation to be too slow, and often feel they are not being given enough support owing to the passive or a reactive

Table 4.3. Stimulating (+) and hindering (-) factors in Dutch regional projects.

Region	Agenda and sustainability	Type of coalition (actors and roles)
Overdiepse Polder	+ Creating possibilities for water storage by building each farm on a *terp*; focus on Planet (water) and Profit (income prospects) + The plan adapts to trends associated with climate change, supporting policy goals such as 'room for the rivers'	+ Co-operation with the local council and the provincial authority + Linked to Habiforum + Stimulation role of individuals within the municipal and provincial authorities
Sjalon Flevoland	+ Focus on Profit, but also some Planet and People aspects; the goal is to expand the scale of agricultural management and increase incomes − Different opinions on the regional image held by the local council and the business group	+ Co-operation between entrepreneurs from different agribusiness sectors + Co-operation with the provincial authority, knowledge institutes and a committed advisor + Linked to TransForum − Attitude of the local council perceived as passive − Critical attitude of the farmers' union, the ex-chairman is also a local executive councillor
NFW Northeast Friesland	+ Focus so far on Planet: high quality small-scale landscape, habitat management and reduced manure production through regional self-regulation; the second phase will pay more attention to Profit	+ Combination of four farmers' associations + Co-operation with NGOs and multilevel governments + Close long-term interactive co-operation with knowledge institute WUR and linkages with TransForum + Temporary agreement with the agriculture minister
New Mixed Business Limburg	+ Focus on Profit (business) and Planet (environmental aspects)	+ Coalition between entrepreneurs, co-operating with the local council, provincial authority and research institutes. Linked to TransForum and adopted by the Agrologistics Platform + Approval from local government for the establishment of the business − Resistance from the local Socialist Party − Role of the media
Regional Innovation Centre Utrecht	+ Focus on Planet (sustainable land use) and Profit (professionalising multifunctional agriculture)	+ Coalition with a broad public-private network at the local, regional and national scales, knowledge institutions and fundraising organisations

Modes of alignment (contacts, rules and personal aspects)	Resources
+ Formal/informal contacts and negotiation behind the scenes + Participation in governmental networks − Rules: general measures, such as the requirement to store manure, do not match the specific situation − Rules: the environmental impact procedures required alternative plans − Increasing differences in interests of farmers	+ Power: use of the media + Power: commitment of the state secretary + Access to knowledge − Time: long duration of the process
+ Leadership and initiative by a farmer with vision − Obstacles encountered by leaseholders in contacts with the State Property Department and the Ministry of Economic Affairs − Difficulties in co-operation due to uncertainty, loss of independence and legal consequences	+ Subsidy from the province for making a business plan − Skills: lack of competences within the local authority − Money: high land prices
+ Strong commitment among farmers − Weak link with the regional economy	+ Knowledge: support from rural sociology researchers, for example on monitoring of sustainability effects + Use of communication and media + Power: direct access to the agriculture minister − Rules: European and national policy on manure obstruct regional management
+ Support of local government − Finding the binding factors between new partners is difficult − Lack of societal and political acceptance; strong protests by citizens against 'megafarms'	+ Support from Knowhouse, a practice-oriented knowledge consultancy + Financial support from the agriculture ministry + Technological innovation − Rules: restrictive spatial and environmental regulations, difficulties with obtaining the necessary permits, and additional sustainable criteria for the final permission − Financial risks − Time: Long duration of the process
+ The initiator is a well-known figure with influence at the national level − Difficulties in aligning different provincial authorities and building a strong business community − The risk of getting too far ahead	+ Money: financial support from national government + Support from researchers at various knowledge institutes − Power: competition between regional organisations

Table 4.3. Continued.

Region	Agenda and sustainability	Type of coalition (actors and roles)
Green Care in Waterland	+ Focus on People (care) and Profit aspects (commercial organisation, source of income for farmers) + Adapting to the trend in new concepts for health and care + Synergy between the strategy and the city branding of Amsterdam as a sustainable city	+ Coalition with the City of Amsterdam and other councils + Linked to TransForum + Linked to the national 'Versatile Countryside' taskforce, supported by the government
Branding Het Groene Woud	+ Focus on Profit (regional economy) and Planet (quality of the landscape)	+ Business community of agricultural and non-agricultural entrepreneurs + Co-operation with local councils and the provincial authority + Direct contacts with Telos, an intermediary knowledge institute – Weak coalition with surrounding cities
New Markets in Heuvelland	+ Focus on Profit, in relation to landscape – Different agendas for the region – Difficulties in combining new markets with landscape quality	+ Co-operation between entrepreneurs in projects + Co-operation with the provincial authority + Network of regional stakeholders + Linked to TransForum (after the first phase) – Historical rural-urban divide between Heuvelland and the city of Maastricht

role of government. In some cases, like Het Groene Woud, regional policy initiatives are fragmented.

2. *Institutional obstacles* caused by a mismatch between general national and European legislation and specific situations or regional characteristics. This is clearly illustrated by the Northern Frisian Woods. General procedures and rules, such as environmental impact assessments or the obligation to build manure storage facilities, often do not provide an adequate response to a situation in transition, as the case of the Overdiepse polder shows. Innovation techniques can clash with current legislation, as in the case of the New Mixed Business. Contracts with semigovernmental organisations, such as the State Property Department (lessor) in the case of Sjalon, can be a legal obstacle.

4. Pathways for sustainable regional development

Modes of alignment (contacts, rules and personal aspects)	Resources
+ Smooth co-operation with chain actors, the local council and education centre due to intensive communication	+ Access to knowledge at different universities − Power: initial reluctance by the farmers' union
+ Social coherence in the region, efforts by volunteers + Passion and leadership shown by the initiating entrepreneur − Difficulties in building bridges between 'red' and 'green' actors − Difficulties with overcoming the rural-urban divide − Fragmentation of regional initiatives and projects	+ Support from researchers + Access to experiences in other regions (Interreg project) + Starting a regional account (Horus Fund)
+ Co-operation between sectors + Combining economic product-market combinations with spatial development + Informal contacts and negotiation behind the scenes − No strong regional business community − Differences between entrepreneurs' logic and governmental logic − Division between 'red' and 'green' policy within the provincial authority	+ Knowledge support − Power: Clashing interests

3. *Clash of different agendas* for the region. It can prove difficult to compile a joint 'storyline' for the region, for example in Heuvelland, due to different public and private interests (see also the in-depth analysis in Chapter 6).

The institutional policy context does not just throw up obstacles. In the Northern Frisian Woods, Sjalon, Het Groene Woud and Heuvelland cases, provincial authorities stimulated part or all of the process by providing subsidies or other support to elements of the private agendas, the New Mixed Far received financial assistance from the national government, and local councils took a positive attitude towards the Overdiepse Polder and Landzijde initiatives.

Success or failure of the coalitions is not only due to regime power. Other factors obstruct private co-operation between private actors, which is complicated by different interests, cultures, motivations and personal factors. In several projects entrepreneurs formed a business community to combine resources and gain power, but they have different interests and motivations, and want to retain their autonomy. In the case of Sjalon it took a long time to align independent entrepreneurs around the concept of a single company. This problem also arose with the New Mixed Farm and they chose to retain the different companies under one 'umbrella concept'.

Co-operating with public actors and forming public-private coalitions is difficult because of their different motivations, roles in the economy, governance mechanisms and organisational cultures (Klijn and Teisman, 2003). According to Noble and Jones (2006), who analysed public-private partnerships, three types of 'distances' influence the process of co-operation. *Autonomy distance* is the reluctance to form a partnership and sacrifice the organisation's sense of autonomy (Cook, 1977: 74) as well as personal control. Private and public actors may believe that their own organisation already possesses the necessary resources or expertise to perform on their own. *Cultural distance* is when initiators are reluctant to engage in the process of searching for a partner organisation in a sector they often know little about, and may harbour preconceived notions. We see this form of distance in the cases where agriculture co-operates with other sectors, such as recreation or the medical sector. *Cautionary distance* is the natural caution people experience when working with any new partner whose credibility, reliability, trustworthiness, competence and integrity are as yet unproved. Governmental actors in particular are often perceived by entrepreneurs as not being trustworthy partners. Two important stimulating factors for co-operation at the regional level are informal networks and the leadership of private actors who can mobilise people around a joint agenda (see also Horlings and Beckers, 2009 and Chapter 5).

4.10 Towards a new regional paradigm in rural-urban regions?

In response to the forces of urbanisation, agricultural developments, new patterns of production and consumption, and new societal demands, many European rural regions are undergoing a transition towards new functions, activities and patterns. Rural researchers have signified this as a transition in the European countryside towards the emergence of a new rural development paradigm (see for example Marsden, 2003; Van der Ploeg and Marsden, 2008). Rural development is a multilevel, multi-actor and multifaceted process (Van der Ploeg *et al.*, 2000). We go into some of the characteristics of this process and analyse the consequences for governance at the European level.

4.10.1 The differentiated European countryside

According to Marsden (1998) four key 'spheres' reflect the diversified or 'differentiated' countryside. They illustrate the degree to which different rural areas are developing contrasting strategies of adjustment and compromise with the increasing scale of production and the wider economy. The spheres reflect different dynamics, which are generated locally and externally, different production and consumption relationships, and different degrees and types of regulation:

- mass food production;
- quality food markets;
- agriculturally related changes;
- rural restructuring.

Despite the uncertainties associated with the common agricultural policy, the growing health concerns of consumers and the potential effects of the liberalisation of global food trade, mass market food production still dominates the rural land base (see also Marsh, 1997). The majority of farmers are increasingly hooked into the vertical food chain which is dominated by the corporate retailers and manufacturers. These sets of interactions and relationships are national and global in character and are subject to technological changes which demand intensive production and economies of scale (Marsden, 1998). The Overdiepse Polder and Sjalon projects illustrate the importance of economies of scale. The New Mixed Business project shows the global vertical links of the food chain driven by technology development. Het Groene Woud targets quality food markets, which can be seen as a reaction to changing consumer preferences and the introduction of quality criteria in food supply chains.

The agricultural changes in recent years, generally defined as agricultural diversification, involve farmers undertaking new activities like nature, water and landscape management, day care and tourism. The Northern Frisian Woods and Green Care Amsterdam projects are examples of this diversification. These developments are highly location specific, based on specific growth in non-local and non-agricultural markets, and require new connections and networks for their development.

In many rural areas the main restructuring of the landscape has little to do with the agricultural sphere. Here extra-agricultural processes, such as housing, exploit the redefined rural resource. The housing market is becoming more regionalised and the spatial patterns of recreational behaviour cross urban boundaries (Mommaas *et al.*, 2000). As the expansion of urban activities and the housing market drives a suburbanisation of work and supply functions, rural areas are becoming part of a field in which urban-rural boundaries are disappearing (Verwijnen and Lehtovuori, 1999); they are no longer outside the city, but lie 'in between' urbanised areas. In these rural-urban regions, urban and rural activities are becoming increasingly interwoven and local planning is more amenable to inward investment and development. An example of restructuring, not explicitly based on agriculture, is the described New Markets approach.

4.10.2 Governance implications for regional development

What are the implications of the diversified countryside for rural policy? As Marsden (1998: 116) states, rural restructuring processes are heavily influenced, both in type and intensity, by the varying institutional and regulatory structures and processes developed in the differentiated countryside. Given the complexity of supply chains and competing development processes in rural areas, the countryside needs to be defined as different kinds of space, as a series of local-non-local network configurations. Any 'integrated' rural policy will therefore have difficulty integrating the different development processes and supply chain links into the differentiated rural spaces. A consequence of this can be that the differentiated countryside requires regionally differentiated development regimes.

An approach to spatial policies that incorporates urban and rural areas can 'attempt to maximise the synergies (for instance production-consumption linkages, value streams) between urban and rural places within a regional context and be more realistic about the degree to which rural areas can capture economic and social value from rural products, services and resource use' (Marsden, 1998: 116). Flemming also mentions the need for more rural-urban incorporation. He argues for more attention to the mutual dependence between the rural and urban economy in rural research, and a shift in focus to more *exogenous* factors, such as the knowledge economy, innovation systems, competence building, regional specialisation and the like (Flemming, 2007). Some Americans describe this as the need for a 'new rural policy' (Drabenstott *et al.*, 2004: 97-104). In its report The New Rural Paradigm, the OECD identifies a new, multisector, place-based approach to rural development which seeks to identify and exploit the varied development potential of rural regions through new industries such as rural tourism, manufacturing and ICT. Such an approach requires the development of new collective governance arrangements to better integrate a broad range of state and non-state actors, horizontally at both the central and local levels, and vertically across all tiers of government. The core of this new rural policy is a closer linkage between the rural and urban economies and recognising the close interplay between rural development and regional development in general (OECD, 2006: 6). Some cases in the Netherlands illustrate this new approach, such as Heuvelland and Het Groene Woud. The OECD report, the Dutch case studies and experiences in regional development in other parts of Europe (see for example Arnason *et al.*, 2009; Horlings and Marsden, 2010; Van der Ploeg and Marsden, 2008) point to the emergence of a new regional paradigm, which can be regarded as a *regionalisation process*, resulting in new linkages between sectors, businesses, producers and consumers and markets (see Table 4.4).

Opportunities for sustainability are found mainly on the regional level, because here the interplay of functions and actors is most dynamic. Sustainable development is difficult to achieve by merely creating vertical linkages (agricultural chains) or horizontal linkages (like the broadening of agriculture with new services). The challenge is to develop novelties by integrating horizontal and vertical innovation.

Table 4.4. Towards a new regional paradigm.

	Old regional paradigm	New regional paradigm
Direction of development	Vertically oriented agricultural development	Vertically and horizontally oriented regional development
Sector	Monosector	Multisector
Functions (nature, agriculture, tourism)	Monofunctional, spatially divided	Multifunctional, spatially integrated
Products and markets	Agricultural mass production Global markets	Valorisation of local assets; regional (niche) products and services
Business relations	Business to business	Business to consumer, links to new actors
Governance	Rural-urban division	Rural-urban integration

4.11 Conclusions

Based on the descriptions of eight cases in the Netherlands, we have analysed four innovative strategies for sustainable development in rural-urban regions in the Netherlands. They can be seen as regional expressions of the scenarios described earlier (see Chapter 2):
1. Increase in farm scale by (a) regional design (b) new organisational arrangements.
2. Innovative decoupling of agriculture from the environment by clustering different agricultural sectors, integrating and reusing resource streams.
3. Multifunctional agriculture by (a) regional environmental management (b) upscaling innovation and knowledge arrangements.
4. Rural-urban development by (a) rural-urban contracts (b) regional branding and (c) the New Markets approach.

A promising perspective for establishing new networks in rural-urban regions are vital coalitions, innovative forms of informal co-operation between private and public actors that create capacity to act in regional processes. In the Dutch cases new coalitions were established and private initiators play a leading role in these projects. The coalitions have some common elements, such as the introduction of new agendas, a focus on informal multilevel contacts with governments, co-operation with knowledge institutes and the use of different forms of power.

How vital are these coalitions? The coalitions create capacity to act to realise their goals (at least in part), but still face the bureaucratic context of regulations and procedures. The current institutional context is *challenged but not changed* by the initiatives; at least, not at the time of writing. Not all the hindering factors relate to regime power, as some problems arise from co-operation between private actors and the types of distances between organisations.

Based on the projects, we distinguished three types of coalitions: business-oriented coalitions, rural-urban alliances and steering coalitions. Important conditions for vitality were found to be: (1) a strong business community (co-operation between private companies) with a common agenda, in which private initiators with passion work as leaders of change; (2) a sense of urgency in the region, creating synergy between private initiatives and public goals and plans, combined with good timing in order to create a 'policy window'; (3) formal/informal multilevel networks, negotiations 'behind the scenes' and the use of power and the knowledge of research institutes. The contribution of these coalitions to sustainability is not evident in all cases and the People dimension of sustainability is underdeveloped. Systematic monitoring of the effectiveness of activities in terms of sustainability within the projects could give more detailed information in the coming years.

The question remains whether vital coalitions have the potential to contribute to the development of new regimes, ultimately leading to the breaking down of instrumental rural regimes and the establishment of new, more culturally-oriented regimes. We see indications for such a change. However, to throw more light on this assumption more detailed analysis of the regional processes is needed and for over a longer period.

References

ZKA Leisure consultants and planners, Urban Unlimited urban and regional planners, University of Tilburg department of leisure studies, 2005. Heerlijkheid Heuvelland; nieuwe markten en allianties voor toerisme in het Heuvelland. Authorised by LIOF: Maastricht, the Netherlands.

Árnason, A., M. Shucksmith and J. Vergunst, 2009. Comparing Rural Development. Continuity and Change in the Countryside of Western Europe. Ashgate Publishing, Aldershot, UK.

Beckers, T., I. Horlings and R. Smeets, 2009. Landschap in Brabant van lijdend naar leidend; Naar een duurzame financiering van het Brabants Landschap. Telos, Tilburg, the Netherlands.

Cook, K., 1977. Exchange and Power in Networks of Interorganizational Relations. Sociological Quarterly 18: 62-82.

De Goede, R.G.M., L. Brussaard and A.D.L. Akkermans, 2003. On-farm impact on cattle slurry manure management on biological quality. NJAS, 51(1-2): 103-134.

Drabenstott, M., N. Novack and S. Weiler, 2004. New governance for a new rural economy: reinventing public and private institutions: a conference summary. Economic Review Q IV: 55-70.

Flemming, J., 2007. Rural-urban linkages and regional development. Paper for the Conference of the European Society for Rural Sociology, August 2007, Hungary.

Hinssen, J.P.P. and F.L.P. Hermans, 2009. Beelden van een grootlandbouwbedrijf; De Sjalon: het bedrijf en het gebied. Telos, Tilburg, the Netherlands.

Horlings, L.G., 1996. Duurzaam boeren met beleid; Innovatiegroepen in de Nederlandse landbouw. Dissertation, Katholieke Universiteit Nijmegen, Nijmegen, the Netherlands.

Horlings, I. and M. Van Mansfeld, 2006. Back to BSIK? Positionpaper Innovatieproject Green Valley, Telos and TransForum, Tilburg, the Netherlands.

Horlings, I., P. Tops, J. Van Ostaaijen and E. Cornelissen, 2006a. Vital coalitions. In: TransForum, The Organisation of Innovation and Transition, Working Papers no. 2, TransForum, Zoetermeer, the Netherlands, pp. 3-44.

Horlings, I., A.J. Bijsterveld and J. Janssen, 2006b. Het Groene Woud is vele landschappen. In: H. Mommaas, K. Nauta, H. Horsten and L. Knippenberg (eds.), Zoeken naar nieuwe wegen, Themaboek Telos 2005, Telos, Tilburg, the Netherlands, pp. 63-75.

Horlings, I. and W. Haarmann, 2007. The soft stuff is the hard stuff; vital coalitions in rural-urban regions, paper for the ESRS conference, August 2007, Hungary.

Horlings, I. and W. Haarmann, 2008. Botsende belangen en vitale verbindingen, positionpaper over het integrated Project Nieuwe Markten en Vitale coalities Heuvelland, In: TransForum, Innovatief praktijkproject Nieuwe Markten en vitale coalities Heuvelland Zuid-Limburg. Workingpaper no. 8, TransForum, Zoetermeer, the Netherlands, pp. 95-174.

Horlings, I., P. Tops and J. Van Ostaaijen, 2009a. Regimes and vital coalitions in rural-urban regions in the Netherlands. In: K. Andersson, M. Lehtola, E. Eklund, P. Salmi (eds.), Beyond the Rural-Urban Divide: Cross-Continental Perspectives on the Differentiated Countryside and its Regulation, Emerald, Bingley, UK, pp. 191-220.

Horlings, I. and T. Beckers, 2009. Leiderschap en bezieling; de mentale dimensie. In: I. Horlings and W. Haarmann (eds.), Afstand en betrokkenheid; perspectieven op duurzame gebiedsontwikkeling. Telos, Tilburg, the Netherlands, pp. 109-133.

Haarmann, W., I. Horlings and G. Derix, 2009b. Please in my backyard, Parc Hoogveld en de opkomst van de integrated care community. TransForum, Zoetermeer, the Netherlands.

Horlings, I. and T. Marsden, 2010. Pathways for sustainable development of European rural regions. BRASS working paper, University of Cardiff, Cardiff, UK.

Klijn, E.H. and G.R. Teisman, 2003. Institutional and Strategic Barriers to Public-Private Partnership: an Analysis of Dutch cases'. Public Money and Management 23(3): 137-146(10).

Kool, A., I. Eijk and H. Blonk, 2008. Nieuw Gemengd Bedrijf, duurzaam en innovatief? Blonk Milieu Advies, Gouda, the Netherlands.

Marsh, J., 1997. A review of agriculture, horticulture and forestry in the UK economy. Office of Science and Technology, HMSO, London, UK.

Marsden, T., 1998. New Rural Territories: Regulating the Differentiated Rural Spaces. Journal of Rural Studies 14(1): 107-117.

Marsden, T., 2003. The condition of rural sustainability. Royal Van Gorcum, Assen, the Netherlands.

Mommaas, H., M. Van den Heuvel and W. Knulst, 2000. De vrijetijdsindustrie in stad en land: een studie naar de markt van belevenissen. SDU, Den Haag, the Netherlands.

Mommaas, H. and J. Janssen, 2008. Towards a synergy between 'content' and 'process' in Dutch spatial planning: the Heuvelland case. Journal of Housing and the Built Environment 23(1): 21-35.

Noble, G. and R. Jones, 2006. The role of boundary-spanning managers in the Establishment of public-private partnerships. Public Administration 84(4): 891-917.

OECD (Organisation for Economic Cooperation and Development), 2006. The New Rural Paradigm: Policies and Governance. OECD, Paris, France.

Pine, J.B. and J.H. Gilmore, 1999. The Experience Economy: Work is Theatre & Every Business a Stage. Business School Press, Cambridge, MA, USA.

Renting, H. and J.D. Van der Ploeg, 2001. Reconnecting Nature, farming and Society: Environmental Co-operatives in the Netherlands as Institutional Arrangements for Creating Coherence. Journal of Environmental Policy & Planning 3(2): 85-102.

Roep, D., I. Horlings and E. Wielinga, 2009. De werkvloer van een Kennisnetwerk Vitaal Platteland; Kennis maken met regionale kennisarrangementen. Rapport 2009-049, LEI-Wageningen, Den Haag, the Netherlands.

Smeets, P., 2009. Expeditie Agroparken; Ontwerpend onderzoek naar metropolitane landbouw en duurzame ontwikkeling. PhD thesis, Wageningen University and Researchcentre, Wageningen, the Netherlands.

Termeer, C.J.A.M., M. Van Lieshout, G.E. Breeman, W.D. Pot, 2009. Politieke besluitvorming over het Landbouwontwikkelingsgebied Witveldweg in de Gemeente Horst aan de Maas. Leerstoelgroep Bestuurskunde,WUR, Wageningen, the Netherlands.

Tops, P., 2007. Regime-verandering in Rotterdam. Hoe een stadsbestuur zichzelf opnieuw uitvond. Uitgeverij Atlas, Amsterdam, the Netherlands.

Tops, P. and F. Hendriks, 2004. Governance as Vital Interaction. Dealing with Ambiguity in Interactive Decisionmaking. Paper presented at the International Conference on Democratic Network Governance, Copenhagen, Denmark.

TransForum, 2007a. TransForum, op weg naar nieuwe kennis. TransForum Expedition June, CD-rom, TransForum, Zoetermeer, the Netherlands.

TransForum, 2007b. Review van het Innovatief praktijkproject Noordelijke Friese Wouden. Working paper no. 6, TransForum, Zoetermeer, the Netherlands.

TransForum, 2008a. Beschrijving innovatieve praktijkprojecten, TransForum, Zoetermeer, the Netherlands.

TransForum, 2008b. Werkplan. TransForum, Zoetermeer, the Netherlands.

Van der Ploeg, J.D., F. Verhoeven, H. Oostindie and J. Groot, 2003. Wat smyt it op: een verkennende analyse van bedrijfseconomische en landbouwkundige gegevens van VEL & VANLA bedrijven, Wageningen: WUR, see also: www.velvanla.nl.

Van der Ploeg, J.D., H. Renting, G. Brunori, K. Knickel, J. Mannion, T. Marsden, K. De Roest, E. Sevilla-Guzmán and F. Ventura, 2000. Rural Development: From Practices and Policies towards Theory. Sociologia Ruralis 40(4): 391-408.

Van der Ploeg, J.D. and T. Marsden (eds.), 2008. Unfolding Webs; The Dynamics of regional rural development. Royal van Gorcum, Assen, the Netherlands.

Van Mansfeld, M. and H. Van der Stoep, 2007. Procesverslag Heuvelland. TransForum, Nieuwe Markten en vitale coalities Heuvelland Zuid-Limburg; innovatief praktijkproject, workingpaper no. 8, TransForum, Zoetermeer, the Netherlands, pp. 19-94.

Van Rooy, P., A. Van Luin and E. Dil, 2006. Nederland boven water; praktijkboek gebiedsontwikkeling, NIROV/VROM. Habiforum, Gouda, the Netherlands.

Van Rooy, P. and R. Slootweg, 2003. Spiegelproject Overdiepse polder; samenvatting van verkenning met advies van bewoners en ondernemers. Habiforum, Gouda, the Netherlands.

Verwijnen, J. and P. Lehtovuori, (eds.), 1999. Creative Cities: Cultural Industries, Urban Development and the Information Society. UIAH Publications, Helsinki, Finland.

WRR (Wetenschappelijke Raad voor het Regeringsbeleid), 1989. Ruimtelijke ontwikkelingspolitiek. SdU Publishers, Den Haag, the Netherlands.

Chapter 5
Vitality and values: the role of leaders of change in regional development

Ina Horlings

5.1 Introduction

Rural areas throughout the developed world are increasingly undergoing a process of rapid transformation. Many societies are having to come to terms with very heterogeneous landscapes in which agriculture is no longer the backbone of the rural economy, with only 10% of the workforce being employed in agriculture. The expansion and upgrading of transport and other infrastructure networks has attracted new permanent investments and workers to rural areas. The resulting easier commuting over longer distances has enabled people to live in rural areas and work in urban centres (OECD, 2006). Rapid transformations also affect the social landscape. Higher education, access to information, low birth rates and mortality, political diversity and ethnic heterogeneity are no longer exclusively urban values, but commonplace in rural areas as well (Wiskerke, 2007).

All these transformations challenge regions to reinvent themselves and develop in new directions. For many an obvious choice is to set out on the path of economic growth and compete with other regions for survival. There is, however, a growing awareness that in the long run regions should embark on a more sustainable development pathway that includes social, ecological and cultural aspects (Millennium Ecosystem Assessment, 2005; OECD, 2006). In practice, however, sustainable regional development is difficult to achieve. The neoliberal discourse has long held that economic growth is good and total control of nature is within our grasp (Dryzek, 2005). Moreover, programmes for regional development often reflect the vested interests of elite groups. People seeking sustainable alternatives are often excluded from powerful networks and decision-making processes.

An alternative perspective is that of vital coalitions. Vital coalitions are forms of self-organisation, often initiated by private entrepreneurs. We argue that within vital coalitions certain people play a crucial and leading role in setting the direction, mobilising others and aligning them around a regional agenda (Horlings, 2008). We therefore explore new ways to make sustainable regional development work in practice that take a people's perspective, because in the end it is people that have to take action for sustainable development. We are particularly interested in how leadership can be provided. In our study of experiences in the Netherlands we have observed new and fascinating forms of 'value-based' leadership that cannot be explained by traditional leadership models. Our goal is to explore alternative models for analysing these new kinds of leadership.

The central question in this chapter, therefore, is: *what is the role of leaders of change in regional development?* After explaining the methodology in Section 5.2, we build an analytical framework based on the literature to get more insight into 'the art of leadership' in regional development (Sections 5.3-5.6). In Section 5.7 we describe four examples of leaders of change in Dutch practice to illustrate their role in coalitions and regional development. In Section 5.8 we draw on all the eight case studies described in the previous chapter to analyse the characteristics of Dutch leaders of change: their driving forces, their tasks, behaviour and roles, the coalitions they build and the strategies they follow in regional co-operation.

5.2 Methodology

The empirical material in this chapter is derived from the eight Dutch cases described in Chapter 4. These cases ensure a range of contextual variation in terms of location, the size of the project, the sectors involved and the goals pursued. In these cases coalitions are formed in which private leaders play an important role (see Table 5.1).

Data were collected over two years from interviews with the initiators of these projects (in some cases the initiators were interviewed more than once), participative observation during meetings and available documentation. The interviews, which were based on open-ended questions, were structured around the following topics:
- the initiator's idea of what sustainability means and the goals set for achieving sustainability;
- personal characteristics, competences, motivation and passion;
- participation in networks and coalitions, and the roles played in regional co-operation;
- stimulating and hindering factors in co-operation and in relations with the institutional context, and the way the initiator dealt with these factors;
- resources (time, money, capacities).

Table 5.1. Empirical cases.

Project	Private organisation	Respondent
Overdiepse Polder (Noord-Brabant)	Overdiepse Polder interest group	Dairy farmer
Sjalon (Flevoland)	Sjalon business partnership	Horticulture grower
Northern Frisian Woods (Friesland)	Northern Frisian Woods Association	Dairy farmer
Industrial Ecology: The New Mixed Business (Limburg)	New Mixed Business partnership	Chicken farmer
New contracts with the city in Waterland (Noord-Holland)	The organisations Landzijde and My Farmer	Director of Landzijde
Arkemheen Eemland Regional Innovation Centre (Utrecht)	Eemland Farm	Owner of the Eemland Farm and dairy farmer
Regional branding of Het Groene Woud (Noord-Brabant)	Regional Cooperative Association	Strawberry farmer
New Markets approach in Heuvelland (Limburg)	Orbis Medical and Healthcare Group	Ex-member of the board

Additional information about the importance of the personal motivation of regional pioneers was gained during a meeting of regional pioneers in the context of the EU Leader+ programme. These pioneers discussed their passions and motivation during a two-day session in May 2007 (see Horlings *et al.*, 2009b). Using the data obtained from these studies and interviews it was possible to answer the following questions about the initiators, or leaders of change, in our eight case studies:

1. What are the personal characteristics of these leaders of change in regional coalitions, including motivation and passion, competences and roles in networks?
2. To what extent did these leaders of change introduce new sustainability issues or regional agendas?
3. What obstacles did they face in regional development processes?
4. Which strategies did they follow in dealing with the stimulating or hindering role of the institutional context and regime power.

Leadership is situational and differs from individual to individual. To illustrate how leaders of change act in the regional context we describe four leaders in this chapter into detail. We chose to describe leaders of change who were trying to link rural and urban development (the last four in Table 5.1) because these cases reflect an innovative scenario that seems to be promising in Dutch rural areas under urban influences, which are found in large parts of the Netherlands. The Dutch Ministry of Housing, Spatial Planning and the Environment has called for examples of new spatial policy concepts based on a rural-urban alignment and alliances between agricultural and non-agricultural entrepreneurs. In this scenario new actors enter the rural policy arena (VROM-raad, 2004). The OECD also takes this line of thinking, based on its observation that a 'new paradigm' is emerging in rural areas. The core of this new approach is the need to create closer links between the rural and urban economies and to integrate rural development with regional development in general (OECD, 2006).

5.3 The role of leadership in sustainable regional development

To lead means to go before or to show the way, to influence or to seduce; to go head of or in advance of, to have the advantage over; to act as leader; to go through or pass; to act as guide. Why should we study such leadership? Why not just focus on the positions people hold and study the formal powers attached to elected positions? Urban regime theory can help us to answer these questions. According to one of its founders a 'regime' can be defined as 'an informal yet relatively stable group with access to institutional resources that enable it to have a sustained role in making governing decisions' (Stone, 1989: 4). The relevance here is that a regime is first and foremost a political coalition that holds specific ideas about problem definitions and solutions and strongly influences the direction of local or regional development. Urban regime theory takes into account the structural influences on urban regimes, but also stresses that local actors can change 'the rules of the game' by their meaningful conduct. A problem would arise if these actors were restricted to government actors, because these cannot create the necessary capacity to act on their own. To be effective, governments must blend their capacities with those of various non-governmental actors (Stoker, 1995: 58). The urban regime theory explains how different configurations of private-private, public-public and public-private coalitions are structured by, and in turn structure, development agendas at the local level (Davies, 2002; Dowding, 2001; Goodwin and Painter, 1997; Lauria, 1997; Stone, 2002). It also explains that the capacity to act is not given, but must be created and actively maintained. The question is not 'who governs' but how to develop the *capacity* to govern. Leadership is essential in this process (Stone, 1989: 229) and requires more than simply holding public office or a position in government (Stone, 1995: 96).

Aimless interaction between actors requires no leadership. Leadership revolves around purpose, and purpose is at the heart of the leader-follower relationship (Stone, 1995: 97). Leadership is a form of power. It can make things happen that otherwise would not take place. Burns proposes 'contribution to change' as a test of leadership (Burns, 1978: 427). We agree with Burns that leadership comprises three essential elements: it is a purposeful activity, it operates interactively with a body of followers, and it is a form of power or causation. Put succinctly, leadership is 'collectively purposeful causation' (Burns, 1978: 434).

Beckers (2008) describes this point in his analysis of leadership in the Eindhoven region in the Netherlands between 1992 and 2002, after the closure of the DAF automobile factories. He defined the situation as a regional regime characterised by co-operation, leadership and trust. A coalition of three individuals representing local government, science and business provided leadership in the form of network power, connecting ideas, inspiration and individuals, and by using windows of opportunity. The coalition developed the concept of 'Brainport Eindhoven' to express the potential for a high-tech region. The leadership of the three initiators depended strongly on the specific regional situation. Success factors in this process included a sense of urgency caused by high unemployment following the closure of the DAF factories, agenda building and telling the story of the brainport with passion, building a coalition between government, the business community and the university, and the mobilisation of resources (money, time, competences) (Beckers, 2008).

In the context of globalising regions, providing leadership is a complex business because a great variety of actors with different interests meet in the regional arena, an arena in which no single governmental body has a monopoly on steering change. For this reason, governments seek to mobilise social capacity at the regional level through horizontal forms of steering and forms of co-production, and by stimulating negotiation between public and private actors. These responses may seem to be promising, but in practice actors often encounter problems related to clashing discourses, unequal power relations, selective choice of actors and a lack of co-operation between interest groups. *The problem is that actors try to achieve sustainable regional development by employing new styles of governance, but still use traditional arrangements.* Regional co-operation is particularly problematic. Regional development requires the involvement of innovative ideas and partners, but public and private actors face the risk becoming incorporated into 'solidified arrangements' that are not suitable for dealing with new challenges (Horlings *et al.*, 2009a).

This problem has been recognised in many OECD countries. In these countries there is a trend towards more decentralised policies, leading to new responsibilities at subnational levels. At the same time, there is an increased focus on the role of local entities in the implementation of such policies and bottom-up approaches are being encouraged in many countries. This throws up new questions. What constitutes the obstacles to co-operation and co-ordination at the local level of rural policy? What are the most effective mechanisms for enabling co-operation between different local actors? In various OECD countries local partnerships face a number of potential obstacles, such as the complexity, rigidity and fragmentation of national and supranational policies, which constrain regional development (OECD, 2006). Local initiatives often succumb to administrative and institutional 'lock-in' caused by rigid procedures, regulations and institutional arrangements.

Recent challenges complicate governmental steering at the regional level even more. First, the distinction between rural and urban areas is becoming increasingly blurred and in many regions the urban-rural dichotomy no longer holds (Wiskerke, 2007). In such 'metropolitan landscapes' urban and rural activities are increasingly interconnected. These areas are being transformed from a domain dominated by agriculture to a domain dominated by the consumption demand of the growing urban middle class, often described in symbolic and cultural terms (nature, authenticity, freedom, etc.). Metropolitan areas have become network societies where local and international production and consumption are connected in a complex system, but policy and policy implementation is still organised along rural and urban sectoral lines.

Second, new and more urban demands are made of these metropolitan landscapes, and in some cases current land uses (urban sprawl, large-scale agriculture) do not reflect societal values like the maintenance of valuable agricultural landscapes. As new functions, new inhabitants and new practices take root in rural areas, the power of rural actors decreases. These changes diversify and complicate the rural arena, raising questions about conceptions of rural space (Frouws, 1998; Marsden, 1999).

Third, new challenges, such as water safety and climate change, cause problems of sustainability. Tackling these challenges requires new alliances between actors from different sectors. To make this even more complicated, the area required for regional development can vary considerably, depending on the social and physical geographic environment, natural resources and amenities, skills and infrastructure (OECD, 2006).

We argue that for sustainable regional development new arrangements are needed that are able to cope with these new challenges and allow the introduction of new discourses, actors and rules, as well as a redistribution of power. Such arrangements cannot be 'designed', but rather evolve over time. This brings us back to the leadership issue because we think leadership is the key to this process. Leadership is not a magic ingredient or an easy solution to complex problems; it is a profession based on individual skills.

Leadership acts mainly through *networks*. It is an aspect of the relations between leaders and their social context and is expressed in the interaction between the driving forces and ambitions of leaders and the needs and attitudes of the collective they are trying to lead. Leadership is a societal function and task to enable co-operation ('t Hart, 1999). The importance of networks is also recognised in natural resource management and nature conservation (Zimmerer, 2006). Here, networks link people and their activities across areas of resource use (e.g. farmers and land use networks) or between rural and urban areas (e.g. food and migrant networks). Such new spatial relations require advanced forms of leadership beyond organisational boundaries.

Leadership is also not matter of leaders and followers, but a collaborative process. Sotarauta (2002) refers to this process as 'shared leadership':

> *In this kind of context, leadership may be seen as the effect of actors on one another and it may be that in the promotion of regional development, there are several leaders with different qualities. At all events, leadership in regional development is a more or less collaborative process. No one can lead the entire region alone since it is not possible to control the activities of the other actors. It is about seizing leadership and how it can be seized.*
>
> (Sotarauta, 2002: 186)

In the remainder of this chapter we explore in more depth the 'secrets' of leadership. In the next two sections we discuss the leadership tasks required for sustainable regional development and specific leadership capabilities.

5.4 Leadership types and tasks

This attention to leadership is not a plea for 'strong leaders', nor is it merely a question of position, status or formal power. Leadership very much depends on the situation, conditions and circumstances. A leader's capacity to act is hindered by impersonal, institutional factors such as laws, rules, culture, policy-making procedures, routines and political power relations (see above). The relevant questions with respect to regional development are what type of leadership is needed to change the regional situation, and what are the regional leader's key tasks?

The huge amount of literature on change management in organisations can be helpful in addressing these questions. The model of dynamic conservation by Schön (1973, 1983), for example, explores the inherently conservative nature of organisations and their resistance to constant change. Schön recognised the need for what he termed the 'learning organisation'. These ideas are further elaborated in his reflection-in-action framework, the mapping of a process for coping with this constant change. There are many more approaches to managing change, such as 'appreciative inquiry' (Cooperrider *et al.*, 2008) scenario planning, 'organise with chaos' (Rowly and Roevens, 1999) and 'theory U' (Scharmer, 2005). These approaches, however, are too limited for achieving sustainable regional development. Leaders have to lead not only within the boundaries of the organisations and communities that authorise them, but also to reach out beyond their field of formal authority and influence other organisations and communities through their words and actions.

Sotarauta (2002: 197-203) identified various types of leaders in regional development practice, each with different roles and qualities: technocrats, network shuttles, visionaries, handicraftsmen, political animals and battering rams. These roles are briefly explained in Box 5.1.

Key tasks in shared leadership in regional development are awareness raising, mobilising, framing, co-ordinating, and visioning between visions. These leader's tasks are described below (see Figure 5.1).

5. Vitality and values

Box 5.1. Types of leadership in regional development (Sotarauta, 2002).

- *Technocrats* create rules, structures and various institutions within established modes of action. In regional development processes they maintain stability and make sure that rules are followed; they are systematic and precise.
- *Network shuttles* are co-operation-oriented, prepared to discuss and look for common interests. They are able to use network power and 'seductive moves' (moves that other players do not have to respond to, but want to because they take into account other players' strategies and goals).
- *Visionaries* have imagination and the ability to see the big picture. They not only shape what activities will be important in the future and how they will be carried out, but also ponder on completely new kinds of activity combinations, or on what kinds of completely new games might emerge.
- *Handicraftsman* pay attention to the needs of the moment, and try to make different processes go as smoothly as possible. They are responsible, balanced and helpful.
- *Political animals* are interested first and foremost in their own positions and in the future. They are chameleons by nature and can be different types as long as the role they select serves their own ends.
- *Battering rams* are very goal-oriented. They use all possible means, from seducing to negotiating to forcing, to get what they want: they talk, convince, envision, network, change the rules of the game.

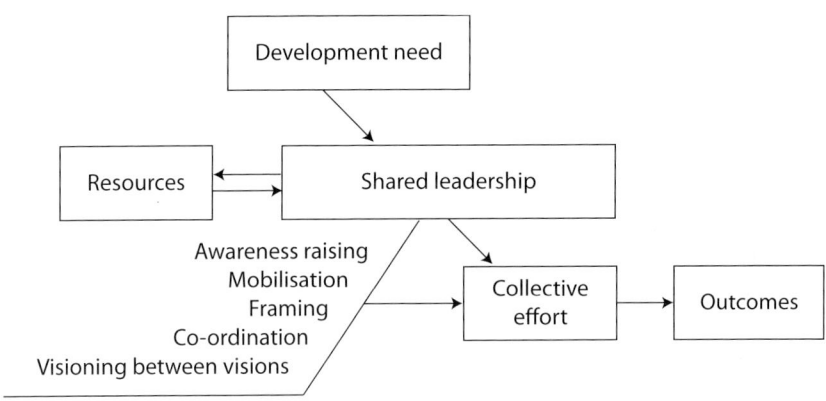

Figure 5.1. Shared leadership and its processes (Sotarauta, 2006).

Awareness raising refers to the task of gaining attention and then directing it to the questions and issues that need to be faced. To do this, one has to provide context for the action and a *story line* that givens meaning to action. Other actors need to comprehend the purpose of adaptive or transformative measures and thus various partners need to be actively involved in the sense-making process (Heifetz, 2003: 225).

Mobilisation is one of the core tasks of leadership. The task is to provide all relevant actors with a seat at the table when strategic issues are framed and strategic decisions are going to be made, and motivate them to take a seat. Mobilisation depends on the willingness of potential participants to devote resources to the network and to be influenced by actors who may have other interests at stake. Seducing individuals and organisations into engaging in the collective efforts, and maintaining this engagement and commitment, is a fragile and subtle process (Sotarauta, 2006).

Framing is about finding a shared perception of issues, a shared vocabulary and shared mental models. Shared mental models are among the strongest glues holding networks together, but the problem is that they emerge slowly and require many face-to-face conversations between network participants to become established. Leaders can introduce new ideas in the network (Kickert *et al.*, 1997). Leaders who are able to define and frame the problems and issues to be addressed in regional development processes are able to exert the greatest power.

Co-ordination is needed to improve communication between fragmented groups of actors, to foster and organise collaboration and to influence the division of labour within the network. It has the power to create and shape institutions or structures. One aspect of co-ordination is identifying and demolishing 'frozen shapes': configurations of administrative structures and government processes that are static and unable to adapt to the changing environment, and thus hamper development work and cause lock-ins (Sotarauta, 2006). In this sense, frozen shapes can refer to regimes that have outlived their time because they have not adapted to changing circumstances. Another part of the co-ordination tasks is to build trust, mutual dependency, loyalty, solidarity and horizontal co-operation based on reciprocal support between organisations and individuals. A third aspect of co-ordination is producing shared and often tacit knowledge that strengthens the social integration of actors in a way that goes far beyond the institutions and networks to which they belong, giving them the ability to network competently and make effective use of informal relations. The ability to share feelings, emotions, experiences and mental models is therefore important (Sotarauta, 2005).

Effective regional development requires networks that can stimulate discussion of visions of a different future and transform these imaginative ideas into focused strategies and action. In the hands of a network leader, a shared vision is a powerful tool for giving direction to a process. In practice, shared visions are combinations of the goals and visions of individual actors. A shared vision is a process that is based on *visioning between visions*, in which appreciating 'other visions' is crucial. Learning about other actors' thinking patterns, and especially about their views and perceptions of futures, is a core skill of the network leader. What often appears as a collective entity is a complex, constantly evolving process between the whole and its parts, between the collective and the individual (Sotarauta, 2006).

5.5. Leadership capabilities

In knowledge-based and continuously changing societies, leaders must continuously use their 'feelers' to probe the reactions and intentions of others, allowing them to learn, innovate and adapt to changing situations. Leaders have to be both dynamic and persistent at the same time.

Regional development needs new kinds of leaders who possess the skills required by the network society and who have an understanding of a new kind of power.
(Sotarauta, 2002: 187)

Leadership is not about attaining ego-driven goals, but adapting to the higher goal of required social change. This requires *servant leadership*, which is open, flexible and participative. Compassionate commitment to a higher cause is an ordering principle for managing interdependent relations in a changing, globalising world. Leadership can create new connections between different sectors of society. It requires the ability to inspire others, to listen and to create trust. When leaders contribute to an effective form of *dialogue*, which goes beyond dominant prejudices, the power of collective thinking can generate new creative perspectives (Jaworski, 1998).

Besides formal power, which is connected to the leader's position in the network, network power is one of the essential latent resources for exerting influence. Network power is based on a set of loose and fixed linkages between networks formed by individuals and organisations. Today's leadership does not advocate forcing moves, but emphasises seductive moves in these networks:

A seductive move is based on the fact that other players do not have to respond to it, but they want to, because it takes into account other players' strategies and goals. While the forcing move attempts to make other players yield to what it wants, the seductive move attempts to make other players co-operate.
(Sotarauta, 2002: 188)

Influence in regional development processes is seductive by nature rather than forcing. In practice, influence builds on different forms of power, but first and foremost on interaction and social skills (Bragg, 1996: 43). The leadership skills required in regional development are the ability to work in a team, interactive influencing, and most importantly, collective sense-making and storytelling (Sotarauta, 2002: 205-206). Sense-making means 'the making of sense' (Weick, 1995: 4), whereby active agents (such as leaders) construct 'sensible, sensable' events (Huber and Daft 1987: 154). Collective sense-making means looking for the meaning of events by creating new knowledge and by recognising, renewing and creating interpretations. Storytelling refers to the ability to visualise thinking, to use metaphors and narratives to get the message across (Weick, 1995: 4). In short, the ability to generate creative tension, excitement and arousal is a key to successful leadership (Sotarauta, 2002: 204). Based on this notion, we can derive a clear set of related abilities. These are summarised in Box 5.2.

5.6 Value-oriented leadership

Applying the leadership capabilities and carrying out the tasks described above does not guarantee success in all circumstances. Shared leadership in networks is highly contingent upon the underlying *cultural aspects* and *values* that influence co-operation (see also Horlings and Padt, 2009). Public and private organisations have different motivations, roles in the economy, governance mechanisms and organisational cultures (Klein and Teisman, 2003). The

> **Box 5.2. Leadership capabilities for sustainable regional development (Sotarauta, 2005).**
>
> - *Institutional* skills: the ability to create an institutional set-up that supports the promotion of the competitiveness of a region and the ability to remove the institutional obstacles and bureaucratic rigidities that block processes and networks.
> - *Networking* skills: the ability to forge trust, mutual dependency, loyalty, solidarity and horizontal co-operation based on trust and reciprocal support between organisations and individuals.
> - *Socialisation* skills: the ability to produce shared and often tacit knowledge that leads to social integration of actors. The ability to share feelings, emotions, experiences and mental models is therefore important.
> - *Absorptive* skills: the ability to identify, assimilate and exploit knowledge from the environment.
> - *Interpretative* skills: the ability to prevent or resolve deadlocks arising in the development process because actors have different, deep-rooted perceptions of problems, which can result in a 'dialogue of the deaf', by maintaining or creating conditions for an open debate. Openness and transparency are essential qualities.
> - *Strategic* skills: the ability to make decisions about the focus of regional development by (a) defining strategy and vision in a collaborative process, (b) translating visions into strategies and actions, (c) changing situations of crisis into constructive situations, (d) starting, managing and leading processes in different phases, (e) good timing to exploit competitive advantages as a pioneer, (f) presenting big goals in such a way that they appeal to others, and (g) taking strategic decisions.
> - *Excitement* skills: the ability to generate and capitalise on creative tensions between the inspiration of key individuals and the dominant thought patterns. It is the ability to get people interested and motivated to participate in development efforts, to create a sense of urgency and drama, and to achieve short-term results to keep people committed.

literature on public-private partnerships indicates that there are people who can bridge the gap between organisations with different cultures. These 'boundary spanners' make sense of their situation by labelling partnerships as a threat or an opportunity (Noble and Jones, 2006: 903) and develop personal and emotional bonds based on mutual trust, commitment and respect (Jones and Noble, 2008: 111). Individual actors attribute values to regional assets and qualities, the social cohesion within the region, and regional identity. In regional development process attention is often focused more on the necessary political, economical and technical changes and less on the changes in the social and natural environment, which are experienced by the region's inhabitants. Identifying these values is particularly important because in regional processes different images, attitudes and perceptions about the region interact (Duffhues, 2009). These values can be a source of inspiration for leaders of change (Horlings *et al.*, 2009b).

Culture and values are largely underestimated in the leadership literature, or only implicitly referred to. In the rest of this chapter we propose a 'value-oriented' leadership model that includes these aspects, but is much broader in scope. We not only want to take into account how leaders can effectively work in networks (see above), but what drives them to act and how they can initiate wider institutional change. To this end, we use the psychological diagram of Wilber (2000) as a starting point (Table 5.2). The upper row of the diagram represents the individual perspective and the bottom row the collective perspective. These perspectives are divided into the subjective inner world and the objective outer world. The resulting

Table 5.2. Wilber's psychological diagram (Wilber, 2000).

	Inner world	Outer world
Individual	I: Intentional (subjective)	IT: Behaviour (objective)
Collective	WE: Cultural (intersubjective)	ITS/THEY: Social (interobjective)

quadrants, or dimensions, are referred to as I (upper left), IT (upper right), WE (lower left) and ITS/THEY (lower right).

We have reinterpreted Wilber's diagram to develop our model of value-based leadership (Table 5.3). The WE dimension refers to the intersubjective perspective, which we link to our earlier notions of shared leadership and associated capabilities, which emphasise the importance of co-operation, coalitions and cultural aspects and values. But a holistic view of value-based leadership also includes the other three dimensions, which we discuss below.

Table 5.3. Wilber's psychological diagram applied to regional development.

	Inner world	Outer world
Individual	I: Personal qualities and driving forces	IT: Behaviour and roles
Collective	WE: Co-operation, coalitions, cultural aspects and values	ITS/THEY: Strategies for dealing with the institutional context

5.6.1 Personal qualities and driving forces (I)

The I quadrant is about what keeps leaders of change motivated and energetic, what keeps them going. We put personal qualities and attitudes, and the underlying personal motivation, passion and drives into this quadrant. According to the management literature, besides facts and rational thinking, 'soft' factors like personal enthusiasm, stimulating motivation and commitment to others are also important aspects of leadership (Buchanan and Boddy, 1992). The most important things a leader can bring to a changing organisation are passion, the ability to convince people and confidence in others (Kanter, 1999). Empirical projects in the Netherlands show that facts, feelings and emotions do indeed play an important role and contribute to informal personal understanding between people (Horlings and Van Mansfeld, 2006). When facing obstacles, inspiration, passion, energy and inner motivations are crucial for building effective coalitions. It is the essence of the 'human factor' in regional projects. Passion, in particular, is an essential quality in enabling participants to persist during long processes of co-operation (Horlings *et al.*, 2009b).

5.6.2 Behaviour and roles (IT)

Leadership in networks can succumb to the effects of voluntarism. As sustainable regional development requires that the 'outer world' has to be changed as well, we need to explore how leadership can bring about wider change. The IT dimension refers to perceptible behaviour and roles in a regional setting and is closely related to the institutional context or the ITS dimension (to be described later).

Regional development takes place in different governance contexts, such as autonomy, competition, hierarchy and self-governance. When regional governance is based on autonomy people interact on the basis of exchange, whereas in a competitive environment they interact on the basis of challenge. In a hierarchical context, people accept discipline and mutual differences, but where there is self-governance people take their own responsibility for the network based on dialogue and equal relationships. In principle, there is no single preferred governance context for sustainable regional development. Each can be appropriate and satisfying as long as they contribute to the identity of the network, the willingness to provide authentic input and a positive attitude. An important condition for co-operation is the creation and maintenance of a *vital space* in these situations where trust, specialisation, division of tasks and creativity can flourish (Wielinga, 2001).

However, each situation has a potential negative or regressive side: autonomy can degenerate into isolation; competition can escalate into power struggles; hierarchy can turn into dominance, with oppressors and oppressed parties; self-governance can become bogged down in 'groupthink'. Regressive patterns can be recognised by a loss of energy, resulting in a network that gets stuck in inertia or dissolves into chaos. Leaders play different roles in regional processes, depending on the tension between positive co-operation and regression in each of the four governance contexts. Wielinga's model (Wielinga, 2005: 65-67, 2007: 27) is helpful for discussing these roles (Figure 5.2).

Wielinga's model distinguishes between leadership *behaviour* and leadership *roles*. Leadership behaviour can be stimulating or antagonistic. To show positive behaviour a leader can take on a role as inspirator, negotiator, mediator or clown (depending on the governance context). In each of these roles leaders try to create and maintain a vital space. These roles are not bound to a formal or legitimated position. Leadership roles displaying antagonistic behaviour mirror the stimulating ones and can be described as the strategist, fighter, prophet and regulator. This behaviour is effective when 'position play' is required to create a certain position in the network and gain respect by using power and resources. In such instances, however, the leader risks escalation or strengthening existing blockages.

5.6.3 Strategies for dealing with the institutional context (ITS/THEY)

The ITS/THEY dimension of leadership refers to the interaction between the leaders and 'their' network within an institutional context. In public-private co-operation, for example, interaction is severely hampered by environmental regulations and procedures, the sector-based internal organisations of governments, lack of co-operation between entrepreneurs, lack of trust, clashing interests, psychological dynamics and the physical and social characteristics

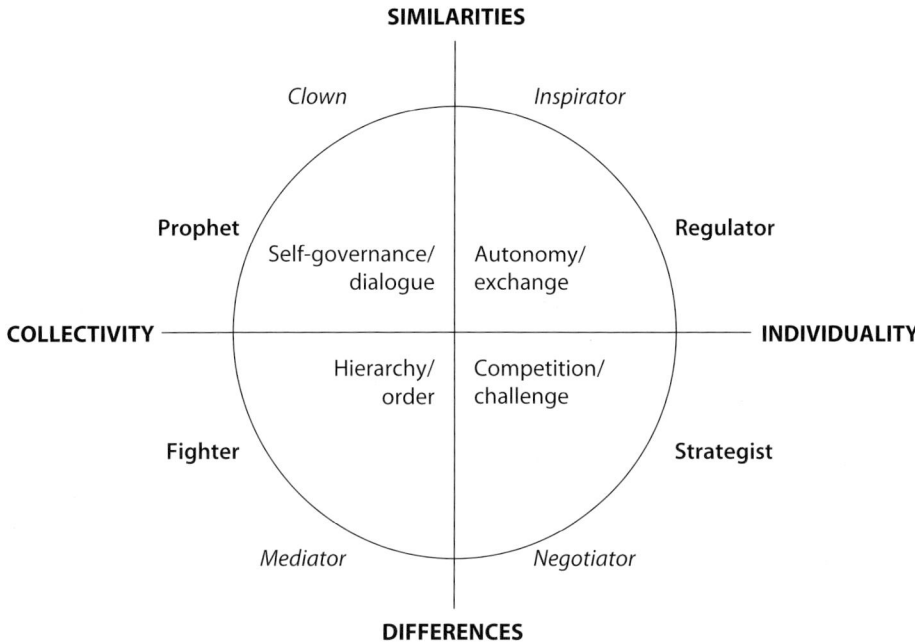

Figure 5.2. Leadership roles (Wielinga, 2007).

of the regional context (Horlings and Van Mansfeld, 2006; Horlings and Haarmann, 2007, 2008; Van Mansfeld and Van der Stoep, 2007). The emotions that result from this, such as despondency, anger, despair and distrust, can easily lead to a lock-in situation.

What strategies do the leaders of change follow to prevent such a lock-in situation? They try to attract new actors to build new coalitions around a joint agenda. A new coalition can contribute to change if it is able to set an appealing agenda and attract powerful and resourceful actors that challenge the existing situation. Often the regional institutional context is protective and conservative towards vested interests and extant power relations. Traditional arrangements will try to hold on to their structure and power. Anyone threatening or affecting these arrangement will face hardship because the institutions retain their power to govern.

In such a situation people are needed who can build bridges between organisations and mobilise actors through dialogue. From their analysis of public-private partnerships, Noble and Jones (2006) distinguished three *types of distances* that influence co-operation: autonomy distance, cultural distance and cautionary distance. In this analysis, leadership is about 'dealing with distances', recognising that these distances exist and trying to build bridges to overcome them. Autonomy distance refers to a reluctance to form a partnership and sacrifice the organisation's 'sense of autonomy' as well as an individual's personal control. If a private or public actor wrongfully believes its organisation already possesses the necessary resources or expertise to act autonomously, this creates autonomy distance. Cultural distance is created when initiators search for partner organisations in a sector that is unknown to them or about which they have preconceived notions. This form of distance can be observed when different sectors co-operate but still use their own culture, language and logic. Cautionary distance originates from the risk of working with any new partner who is 'unproved' in terms of credibility, reliability, trustworthiness, competence and integrity. Governmental actors are often perceived by entrepreneurs as not being trustworthy partners.

Leaders of change seek to bridge the distance between their own innovative network and the existing regime. They work as 'boundary spanners' in public-private co-operation (Jones and Noble, 2008; Noble and Jones, 2006) and constantly switch between networks and related cultures, logics and languages. Various innovation studies have shown that radical innovations often come from outside the existing institutions and are initially developed by private entrepreneurs. Van de Poel calls these actors 'outsiders' (Van de Poel 2000: 384). Elzen *et al.* (2008) use the term 'hybrid actors', which reflects the fact that they operate between the 'insiders' and 'outsiders' of regimes. These hybrid actors stimulate the *anchoring* of the relation between niche and regime. Being too closely associated with the current institutions increases the risk of being incorporated into these institutions, with a consequent loss of vitality (Bang, 2004).

Padt (2007) is on the same track when referring to 'institutional bricolage'. Bricolage is derived from the French verb *bricoler*, meaning to tinker or to fiddle, which comes close to the English 'do-it-yourself'. Following Douglas (1986), Scott (2001) applied the concept of bricolage to institutions, which is interesting because it expresses the idea that institutions do not just impose constraints, but can present opportunities as well. Scott (2001) states that institutions:

> *... are also enabling to the extent that they provide a repertoire of already existing institutional principles (e.g., models, analogies, conventions, concepts) that actors use to create new solutions in ways that lead to evolutionary change.*
> (Scott, 2001: 193)

Padt has investigated regional environmental planning in the Netherlands since the early 1970s. During this period the three government ministries responsible for regional environmental planning developed and institutionalised contrasting policy arrangements. The department responsible for water management established an authoritative arrangement, the environmental department an arrangement based on negotiation, and the agriculture department an arrangement based on self-governance. Juxtaposition of these contrasting policy arrangements caused a lot of frustration and anger at the regional level because in concrete projects the three government departments followed their own individual policy arrangements. Despite this unfavourable context, though, some regions were quite successful because of the presence of regional leaders (see also Padt, 2006). These leaders were able to link the three arrangements in a practical and creative manner, complying with national legislation and provincial planning policies, while at the same time creating room to manoeuvre to establish surprising coalitions and innovative projects. The key to their approach was using the departmental arrangements as an opportunity rather than a constraint. In this sense, they applied a form of institutional bricolage. In Section 5.7 we describe four examples of leaders of change for rural-urban integration in regional development.

5.7 Leaders of change in regional projects in the Netherlands

5.7.1 A farmer in town

Jaap Hoek Spaans runs a farm near the city of Amsterdam. 'Modern, multifunctional farming, taking the market into consideration, offers solutions for urban problems,' he says. In 2000 he launched Landzijde, an organisation which co-ordinates the placement of clients at 87 farms that provide day care. The farmers offer a broad variety of daily activities to their clients, providing safety, structure, space and challenges. The care programmes also help the clients to develop relations with each other and with the animals and nature. Working outside contributes to a better physical and mental health. 'One does not have to say much, but has to do a lot,' says Hoek Spaans.

His personal goal and motivation is to establish new rural-urban coalitions: 'Farmers have to offer citizens recreation, care and food. We have to give new meaning to the area'. Hoek Spaans develops networks of people who want to work professionally: 'I believe strongly in regional networks that are specialised in aspects of multifunctional agriculture'. His passion and enthusiasm are evident; he prefers to see opportunities instead of obstacles and communicates his message with persistence. He also set up My Farm, an organisation which stimulates the development of regional products. My Farm co-operates with Amsterdam on developing a more sustainable food strategy. One of the outcomes of this was the opening of a co-operative supermarket in the centre of Amsterdam in February 2007. This shop sells quality products from the farmers in the Waterland region and functions as a marketplace, so no storage space is needed. Landzijde is also successful. Hoek Spaans encourages local councils to co-operate with Landzijde by communicating intensively with them. Amsterdam contracted Landzijde to offer farm activities to unemployed residents of the city. He also co-operates with universities in the Green Care project by investigating the medical results of on-farm health services, and has worked with schools to develop an educational programme on farms. He started Landzijde as a commercial organisation without volunteers and places great importance on client satisfaction: 'Heart and soul have to be committed to the client'. The board of the organisation consists of non-farmers: 'Progress is made with people who have the insight, vision and entrepreneurial passion for the whole. I learned a lot from people working outside agriculture.'

Besides being a networker, Jaap Hoek Spaans can be described as a 'battering ram' because of his perseverance in working across boundaries. In the beginning he encountered resistance from the farmers' union, which usually works on a project-by-project basis. His opinion is that entrepreneurial innovation may stop after the end of a formal project if the period of assistance is too short. Landzijde offers ongoing help to farmers on bureaucratic procedures and provides advice and expertise on a permanent basis. He realises the danger of an organisation being dependent on the leader. 'One of the dangers is that the publicity focuses too much on one person,' he says, and so he has already appointed his successor. As a pioneer, he does not consider himself the best person to lead an established organisation.

5.7.2 A health manager as project developer

Guus Broos is a leader of change in the province of Limburg and was a member of the management board of Orbis Medical and Healthcare Group in Limburg. He was one of the creative initiators of the new 'hospital of the 21st century' in Sittard (Derix, 2003), and participated in the project *Heerlijkheid Heuvelland* (see also Chapter 6) and a project aimed at developing integral care communities in the region. He can be characterised as a visionary type of leader, with a *vision* for the future of medical care. In his view, the care sector will be too expensive in the near future and must become more patient-oriented and less patronising towards clients: 'We have to create healing environments with an emphasis on wellness. This implies the dismantling of traditional care structures and building the delivery of care around the needs of the client, rather than the production of care.'

As a sociologist who has worked in a broad variety of organisations, Broos is able to build bridges between actors in different settings, both at the strategic and operational levels. He is aware of the benefits of operating in both a formal and an informal network: 'I operate in two networks. First, my informal network contains free spirits like architects, people from religion, business and spirituality. These are people who want to stick their neck out. Second, my formal network consists of people from government, trade and industry associations, businesses and NGOs. They don't feel the need innovate. Some people belong to both networks.' Personal contacts play an important role in his activities. His personal empathy with the director of Castle Hotels in Limburg underpinned the development of the innovative Healing Hills concept, an alliance between the care and hotel sectors. As a follow-up, Broos developed the concept of integral care communities, an urban example being Parc Hoogveld in Sittard. This initiative started as a care centre for elderly people, but developed into a multifunctional service centre for the city community, including the development of a large park. His motivation is to create healthy environments: 'Care is now stigmatising. When people grow old and need services and a housing landscape, they want a trustworthy environment with advanced care and cure' (Derix, 2003). The development and exploitation of the park is funded mainly by private partners, such as housing corporations and investors, who want to create an attractive meeting space for clients and citizens. Broos's networking capabilities, charisma, persuasive skills and status as a member of the largest care organisation in the region were crucial in getting these private developers to support the project and co-operate in its development (Horlings and Haarmann, 2009). His plan is to develop not only urban centres, but also small-scale communities on farms that integrate housing, care and recreational functions. For this, new alliances are necessary. In Broos's view, actors who deal with the content (caring), infrastructure (location and spatial procedures) and finance (investments) have to work with entrepreneurs in the development and exploitation of these small-scale care communities in the countryside (see Haarmann et al., 2009).

Broos is able to influence decision makers at local and provincial level and creates synergy between initiatives, for example by participating in the regional branding process in South Limburg. He is driven not only by a desire to develop new projects, but also to stimulate social cohesion, new communities and voluntary work. 'Parc Hoogveld is based on the concept of the community; it is an urban community. The goal is to realise a housing, employment

5. Vitality and values

and recreational environment that creates social cohesion and a sense of belonging and commitment by combining public and private functions.'

This leader of change was confronted with the high costs of building the new hospital. His position came under pressure during the financial crisis in 2008 and he had to step down. Orbis focused on its core business and jettisoned all its innovative and pioneering activities.

5.7.3 A strawberry entrepreneur in Het Groene Woud

The National Landscape Het Groene Woud is situated in the triangle between the cities of Den Bosch, Tilburg and Eindhoven. In this region, Frans van Beerendonk, a strawberry producer, took the initiative to start a regional branding process with the aim of marketing the key values of the region (Horlings *et al.*, 2006). As a member of the Sustainable Meierij Innovation Group, he encouraged the development of activities for learning from other regions in the Netherlands and Europe. His ideas emerged during an inspiring visit to the Cork region in Ireland where public and private partners market regional products ranging from home-made food to four-star hotels.

Van Beerendonk's passion is rooted in his commitment to his own region: 'I want the region to remain attractive for its inhabitants and for visitors. I want people to become reconnected to their region, so that they can talk about the region with pleasure.' The entrepreneur can be seen as a network shuttle. He has a clear message, which is promoting networking with all kinds of partners inside and outside the region and stimulating people to work together on a voluntary basis for the region. His goal is to develop a regional economy that contributes to the quality of the valuable small-scale landscape. Crucial to realising this goal is getting the commitment of the three cities and non-agricultural businesses: 'The big cities are neglecting their backyard. Businesses that benefit from the countryside should contribute to sustainability. This is how to develop a regional economy'. He believes the development of the region should be based on regional characteristics and values, such as the environmental qualities of the diverse landscapes, the current variety of economic activities, and the social cohesion and co-operation between the people living in the region.

Van Beerendonk is the initiator and chairman of the annual regional Green Forest Festival, which is organised by many local volunteers and receives about 20,000 visitors each year. Het Groene Woud is a dynamic region where many different activities take place and many projects are carried out, but it has led to organisational fragmentation, a pitfall for the regional process. Consequently, Van Beerendonk spends much energy on aligning people and organisations around a joint agenda. In his vision, the region can only be promoted if entrepreneurs, recreational organisations, NGOs and landscape and nature conservation organisations all work together.

5.7.4 A farmer-philosopher aligning city and countryside

The Eemland Farm is a demonstration business for multifunctional farming started by Jan Huijgen, an entrepreneur who sees himself as a farmer, philosopher and lecturer. He studied at Wageningen Agricultural University (now Wageningen UR) and philosophy at the University of Amsterdam, where he was introduced in the ideas of Heidegger, Habermas and Lyotard: 'The distance between farmers and civilians created an unpleasant feeling. Perhaps that's why I studied philosophy. I learnt about the idea that there is an emptiness behind the technical control mechanisms in current society; people are not rooted and need a new foundation, a sense of belonging to a place that feels like home.' He also studied environmental philosophy and theology for a while and was a lecturer at an agricultural school, but science turned out to be not the right path for him: 'I always sought the frontiers. I was too curious and hard-headed to become an ordinary farmer, but also too much involved in agricultural practice to be a philosopher. I didn't have the patience for writing books. I have too much passion and energy.' He realised that his strength does not lie in doing one job, but in combining different roles. His decision to diversify the activities on his farm meant he had to let go of his expectations and desires following a period of crisis: 'I had to sell some land, but when you take away a farmer's land you take his soul. Traditional farming means developing the farm, not selling the land. After this horrible process, everything fell into place. The minister started to talk about rural renewal. I had waited for years for that moment'.

The Eemland Farm is a 'farm plus', with a wide variety of activities. It functions as an incubator for new rural concepts and ideas on different scales. Huijgen is a very clear example of a network shuttle because he connects networks on different geographical scales. For example, he started an association for farmland conservation management in the National Landscape. On a higher level, he founded a rural-urban co-operation network with other associations to stimulate multifunctional agriculture in the Netherlands. Eemland Farm became a regional innovation centre in March 2008.

Huijgen was awarded the 2007 Sicco Mansholt prize for his dedication to multifunctional agriculture and rural-urban alignment. He wants, in his own words, to 'stimulate meetings between farmers and citizens, between the city and the countryside and between creator and creation.' In 2007 he was the driving force behind the conference 'City Seeks Farmer', which was held at Eemland Farm. During this conference the Amersfoort Agreement *(Akkoord van Amersfoort)* was presented to the Minister of Agriculture, Nature and Food Quality. The goal of this voluntary agreement is to strengthen the relation between the built-up area and the green rural environment by developing concrete local projects.

This pioneer's passion is rooted in his agricultural and Christian background. 'I have never renounced my agricultural roots. During my education I always had the feeling that I have to maintain this connection with the love for nature, God's creation, farmer's culture. As a Christian I feel responsible for a sustainable world and caring for God's creation.' He believes that farming is a way of living life as it is intended by God. 'You are in direct contact with nature and the greatness of God's creation. As a result you look at society from the right perspective.' He is a philosopher with a mission who sees farmers as managers of the landscape: 'In the agricultural tradition you see the richness of the relations with the elements, the biological rhythm and the care for people and animals. These standards are of great importance in our society, which is drifting. I see it as my task to bring people into contact with the farm, nature and the way food is produced.'

5.8 Empirical findings: a typology of leaders of change in Dutch regional projects

5.8.1 Personal characteristics and inner motives (I dimension)

Every leader of change is different, adapting in their own way to variable circumstances in different regions. Nevertheless all the pioneers share some common characteristics. The respondents are all *men* and at least *40 or 50 years old*. They are in a phase where they have the experience and the time to do this work, and are able to organise their own business in such a way that there is time for innovation (for example, most farmers have a successor who runs the farm). They have a pro-active, flexible attitude, can motivate people and are creative in achieving their goals.

The role of leader of change requires *perseverance, self-will* and *energy*. To keep going during regional processes they also have to be goal-oriented and patient. An example is the Overdiepse polder case described in the previous chapter, in which farmers developed their own plan for their region. Although the government authorities were positive about the farmers' plan, the initiator had gone through 10 years of negotiations and still no start had been made with implementing the plan because of its complexity and the various formal procedures involved. Another example is the process in Het Groene Woud. The alignment of people around a regional agenda requires networking with a large variety of public and private actors, which can be exhausting. This is illustrated by the words of the initiator of the branding process: 'I make people enthusiastic, sometimes too diligently. My network is large, but I put in a lot of energy. It is not really justifiable. I devote too much time to this, nearly all my time in fact.'

To be successful, *self reflection, an open mind for new knowledge* and *the use of variable multilevel networks* is important, according to the respondents. The leaders reflect on their actions, make use of new ideas and knowledge outside their own group and relate to actors on different governmental levels. Self-reflection and getting feedback were mentioned as important to maintain a clear view of the personal development path and to get fresh ideas. In this process of self-reflection pioneers also face their own personal obstacles. As one of them said: 'The resistance is in me. It is about finding my own strength. My inner ego hampers contact with others. I can be overpowering and hurried, which is a risk when making contacts. It is about engaging with the established regime and revealing your vulnerability to show you are being open.'

Leaders of change are able to *link their own private interests to collective goals and future perspectives for their region*. The agendas are rooted in, but also go beyond, individual personal and business goals. In this sense, the initiators function as ambassadors of these collective goals. One of the motives behind the New Mixed Business project, for example, is the search for an answer to the trend of globalisation within the intensive dairy sector. A driving force behind the actions of the Northern Frisian Woods Association is to develop a regional model instead of a business management model in order to realise environmental goals. The goal in Het Groene Woud is to brand the region, integrating social, ecological and economic aspects.

A general perception of the leaders of change is that the inner dimension, personal motivation, values, feelings and emotions are important in regional processes, but are not appreciated enough by decision makers: 'Local people have the knowledge and experience about the past and present of their region and want to commit with heart and soul to their own environment.' The initiators in the eight Dutch cases have a clear inner drive and translate their inner motivation into practical actions in the outside world. The motivation of these change leaders is very diverse, and personal values are as important as ideas about the future. The initiators mentioned various inner motives. One of these is *awareness* of environmental problems and the ethical aspects of animal welfare. Others mentioned their drive for *corporate social responsibility* and a commitment to sustainable production. A third set of motives is rooted in the *values of the countryside* and the desire to maintain the quality of the landscape. Others feel inspired to *connect* with other people. More *egocentric motives* also play a role; some leaders like being pioneers and innovators, being in the spotlights and having influence. Finally, *spiritual or religious motives* are important for some.

However, pursuing personal motives can lead to resistance if a leader gets too far ahead of the pack. As one of the entrepreneurs said: 'In the harsh light of reality it is difficult to stay on track, and there is a risk of becoming isolated'. Inspiration can create tension between the personal ambition and ego of the leader and the desire to realise societal goals. In most situations, tensions arise between innovative pioneers and governments. Besides rational aspects, factors like fear, lack of imagination, lack of trust and differences in power also play a role. Tensions arise in situations where leaders of change want to speed up the process, want to realise different goals or pursue goals in a different way, or want to take over government tasks, such as the management of national landscapes.

5.8.2 Types of leadership, tasks, competences and skills (WE dimension)

Types of leadership

The initiators in our case studies display mainly a combination of two types of leadership characteristics – visionaries and network shuttles – but also show some characteristics of the battering ram. The shadow side of visionaries is that they get bored with details, may seem superficial and impatient, or are seen as people who think they know better than everyone else. They raise new issues even before older ideas have been thoroughly examined, let alone carried out. They are often not prepared or able to carry these ideas out. The most obvious visionaries sometimes find themselves in a position of losing touch with their constituency, neglecting details (sometimes with major consequences) and showing little interest in implementation, with the risk of generating resistance within their organisation or region. An example is the care manager who eventually had to step down due to financial problems within his organisation. Some respondents mentioned the risk of losing touch with other entrepreneurs.

Tasks

The leaders of change raise awareness about their goals, such as a more regional steering model (Northern Frisian Woods), a process of regional branding (Het Groene Woud), integral community building (New Markets in Limburg) or a new form of sustainable intensive farming

(New Mixed Business). An important condition for awareness raising is mobilising people around an issue. We identified four types of mobilisation:
1. mobilising entrepreneurs to form a business community;
2. mobilising policy actors around a joint agenda to frame new perspectives for the region;
3. mobilising volunteers and professionals to organise events;
4. mobilise knowledge workers to strengthen their own insights and knowledge.

To what extent do the leaders carry out co-ordination tasks? The coalitions seem to be reluctant to form new institutions or organisations. The Northern Frisian Woods Association wants to involve the existing steering group of private actors, NGOs and governments in managing the National Landscape. The Overdiepse polder interest group is involved in government consultation groups so that they can influence decisions directly. However, sometimes new organisations have to be created to align businesses. An example is the creation of a commercial organisation for regional branding in Het Groene Woud, which mobilises private actors around the goal of producing regional products. Another example is the Regional Innovation Centre at the Eemland Farm, which was recently founded to reduce the lines of communication between regional actors and knowledge institutes. This reluctance to abandon existing structures entails the risk of incorporation into the institutional context and the leaders are aware of the tension between their situational and institutional logic (see Chapter 3 for an explanation of these concepts). One of the respondents expressed the danger of adapting too much to government timetables: 'I have become more flexible. I find it normal now that things take more time. This is also a risk. People often ask me when we can build new animal sheds. I fight against delays less than I used to.'

Competences and skills

The respondents were asked which competences they use in their daily work, and what they consider to be their strengths and weaknesses; the results were compared with the leadership capabilities described above.

Regarding *institutional* skills, the initiators in the Dutch cases are mostly pragmatic. Their goal is to stimulate flexible decision making, gain influence and establish short lines of communication with organisations. The focus is more on using networks than building new institutions.

Networking skills are very important, especially in the Dutch situation where government planning is fragmented and regional issues involve a complex web of different interests and organisations. All the respondents communicate intensely with a wide variety of public servants, repeating their message time and again. The most successful leaders are able to connect different networks and scale up their activities to larger networks. A big problem is the time-consuming process of convincing governments of their ideas as the responsibility of aligning public actors often rests on the shoulders of entrepreneurs. An example is the Green Valley project, in which the Eemland Farm respondent was involved. An evaluation showed the difficulties and time-consuming process of aligning three provinces with different procedures and financial instruments around a joint goal (Horlings and Van Mansfeld 2006). A complication factor is the relation between public servants and decision makers. One of the respondents noted: 'A difficult aspect is the power play and the differences between public servants and decision makers. The mayor of X and the member of the provincial executive of province Y both want to be in control.'

Socialisation skills are revealed in the way the leaders of change play a stimulating role in the joint development of knowledge, participating in projects run by knowledge organisations like TransForum and Habiforum. They also try to share facts as well as emotions: 'Tell the personal story as well; examine your own heart and feelings. Identify the joint emotions and have the courage to connect with your own weaknesses.' Informal relations play a crucial role within the leader's own group and in coalitions, which requires sensitivity. As the respondent in the Sjalon project stated: 'You have to speak and feel clearly, communicate intensely, but also read between the lines when listening to what others say.'

Leaders of change show *absorptive* skills in the way they translate sometimes abstract ideas and visions, knowledge and technical know-how to their own practice. The respondents are very pragmatic, but are able to look beyond their own interest or company: 'To be able to brainstorm freely you have to be able to let go of your own preconceptions and environment.'

The respondents have *interpretative* skills in the sense that they are open to discussion and knowledge from outside. But some respondents also mention the risks attached to their straightforward attitude and clinging to their own concepts and images: 'I would like to learn to debate more freely with an open mind, but my own vision is holding me back. I want to convince people.'

The respondents consider themselves to be *strategic* thinkers, trying to influence decision making at the right moment. Some of them said they noticed a tendency to express new visions and ideas that are too radical for others to take on board, and the associated danger of decreasing commitment from other entrepreneurs: 'Pioneer isolate themselves; they initiate something new, but can be ahead of their time.'

Excitement skills are expressed by some but seems to be an undervalued skill. Some respondents try to create excitement by organising events, conferences or regional festivals in order to mobilise people and stimulate commitment with the region. Some have direct contact with the media, others give lectures on a regular basis to express their vision. Motivating people is what they do best.

5.8.3 Roles in networks and behaviour in forming coalitions (IT dimension)

Roles in networks

All the respondents that were interviewed mentioned that they act in different roles, including both stimulating and antagonistic roles. The clown and regulator roles were mentioned as being the least important. The initiators see themselves in the first place as inspiring, and the outside world recognises this role. The respondents have strong opinions, which they express with enthusiasm and perseverance to put new issues on the agenda. Concepts like the integral care community, rural-urban contracts, the Landzijde model and regional branding are strongly advocated by the initiators. At the same time they try to maintain their own autonomy, mostly working in situations that can be characterised as 'autonomous' and 'self-governing'. When autonomy tends towards isolation, an inspiring leader can project new ideas or insights, or take initiatives that are appealing to others. When self governance regresses into groupthinking, the prophet can shake people free of their illusion of security. Most respondents recognise themselves in the role of prophet, but prefer the term visionary. They also recognise the danger of losing touch with their own group.

Forming coalitions

How do leaders of change establish coalitions and what are the important conditions for this? A good starting point for forging coalitions with other actors is to form a solid *business community* of entrepreneurs with a common goal and agenda. Establishing such a community is not easy because of the inevitable differences in interests and visions, a lack of trust and loss of autonomy, as well as communication problems. The respondents function as leaders of these business communities. An effective method of 'seducing' a group of entrepreneurs into a process of joint visioning is to form a 'community of practice'. In the Sjalon project the entrepreneurs invented this way of working themselves. Often the business community has an open structure. Entrepreneurs can join the group, but only if they accept the group's agenda. The extent to which entrepreneurial drives can be combined with group interest is crucial for the success of the group. Maintaining the autonomy of individual group members is also important (farmers are particularly attached to their freedom and autonomy). An example is the New Mixed Business project, in which the business units retain their independence within the company. The entrepreneurs in the Sjalon project work together within a new organisational structure, but retain ownership of their land.

The initiators stressed that to establish a vital coalition that can create capacity to act it is important that *government officials play a supporting role*. Realising vital coalitions requires political power and the respondents agreed that effective public-private co-operation requires courage and vision from government: 'An important condition for regional development is that the government authorities in question show courage and vision. It is about feeling, vision and having the courage to pursue the desired course of development.' Some respondents argue that the stimulating role of government should take place outside the formal structures of provincial government to prevent fragmentation. The initiator of the branding process in Het Groene Woud argues for an effective implementation authority for Het Groene Woud that is independent of the provincial government departments and can co-ordinate projects and allocate funds. 'This process should be run from a programme that is not part of the administrative structure of provincial government. Now it takes too long. What is needed is a regional organisation that has control over a programme budget. This would requires a steering and implementing regional authority that evaluates plans.'

All the respondents operate in a way in which *reflection, regular communication* and *the use of formal and informal rules* on how to work together are important characteristics. Some initiatives organise reflection meetings on a regular basis, which enables the participants to share their feelings and expectations, explain their roles and build trust and commitment. One of the initiators formulated described this as follows: 'I thought, if we put our heads together, we would be able to work something out and develop a vision. Therefore I looked for people of intellectual substance who were also compatible. You have to trust each other. We spent a whole afternoon deciding on our individual roles.'

5.8.4 Strategies for interacting with the institutional context (ITS/THEY dimension)

When trying to introduce alternative pathways to the future, leaders of change experience regime power as a glass ceiling of institutions, rules and dominant ideas. They also encounter the tension between different types of logic. Change leaders follow several strategies to deal with the institutional context and create capacity to act (see also Horlings, 2009):

1. *Raising awareness, framing and visioning between visions.* The leaders of change raise awareness and articulate new goals, creating a sense of direction and contributing to the framing of problems by telling their story over and over again. They 'vision between visions' to align people around a joint regional story line. The story line may be a vision of the future but can also be limited to a message or slogan. Its success depends on the ability of leaders to express their vision clearly, their competences in mobilising people, the fit with the interests of other stakeholders and the extent to which the message is appealing to others, especially people with formal positions and power.
2. *Bricolage: balancing between different worlds.* Leaders of change operate in networks by balancing between their own innovative network and the existing regime. In this sense they work as 'boundary spanners' in public-private co-operation. They also recognise the danger of incorporation. One of the entrepreneurs expressed this as follows: 'If the process becomes too institutionalised, you lose your freedom.'
3. *Open innovation.* In their search for knowledge, leaders of change look beyond their own network, a practice referred to as 'open innovation' (Chesbrough, 2003). The initiatives described in this book formed coalitions with knowledge institutes. They also tried to learn from other regions, adapting lessons learned in other regions to their own regional contexts. As one of the respondents said: 'A learning region is receptive to experiences in other regions and at higher geographical scales.'
4. *Creating informal and multilevel networks.* Leaders of change use different skills and competences to mobilise people, especially network power and informal co-operation. They sometimes establish large multilevel networks and are able to move effortlessly between the different worlds of policy, practice and science. The change leaders look for ways to 'change the rules of the game'. When they feel obstructed by local civil servants, they try to find supporters for their plans at different and sometimes higher levels of governance. Such a *leapfrog strategy* can be helpful when the goal is to introduce new innovative agendas or rules, when there is a need for more room to manoeuvre or additional finance or other resources. *Informal personal contacts* are important for leaders of change. They use their position, status or personal competences to exert influence through such informal contacts: 'With eight people you can change the world. Informal networks and key players are important,' one of the leaders said. They create opportunities for negotiations 'behind the scenes', in which it is possible to speak freely, to get a feeling for risks and sensitivities, to influence people and to put new topics on the agenda, especially when these negotiations involve people with network power. 'Many things got done during informal chats outside meetings. I know a lot of people. I can get attention. I phoned the decision maker on

Friday night. He came immediately on Saturday morning to our board meetings.' In the end, however, a formal status and a formal decision is needed to obtain commitment at the political level. A specific example of using network power is to make use of the *media*. In difficult situations where the danger of a deadlock is ever present, one strategy is to use the media to exert political pressure. An example from the Overdiepse polder: 'I phoned X and said that if the plan fails the news would be on television, and the failure would be attributed to bureaucratic inertia and the laziness of the government officers.'

5.9 Conclusions: towards value-oriented leadership

Using the I-WE-IT-ITS/THEY diagram we have explored the dimensions and aspects of leadership that are important for regional development. Table 5.4 summarises our findings.

We think Table 5.4 provides a useful framework for analysing value-based leadership for regional development in relation to sustainability and globalisation. By reinterpreting Ken Wilber's psychological diagram and reviewing the rich literature on related aspects we were able to 'demystify' leadership and to compile a set of operational criteria. However, much work still has to be done in order to assess the quality of leadership in different situations and fully grasp the relations between the I, WE, IT and ITS/THEY dimensions. The I dimension in sustainable regional development studies is still neglected. We therefore suggest more research on this dimension of sustainable regional development, focusing on the passion, values and motivations of leaders of change.

Based on the results presented in previous sections, we conclude that leadership in regional rural-urban networks requires special skills, competences and roles. The goal is to organise a

Table 5.4. Dimensions and aspects of value-based leadership for regional development.

	Inner world	Outer world
Individual	Personal qualities Inspiration Motivation and commitment Passion Emotions	Leadership behaviour (stimulating or antagonistic) Leadership roles (inspirator/regulator, negotiator/strategist, mediator/fighter, clown/prophet)
Collective	Leadership types (technocrats, network shuttles, visionaries, handicraftsmen, political animals, battering rams) Leadership tasks (awareness raising, mobilisation, framing, co-ordination, visioning between visions) Leadership capabilities (institutional, networking, socialisation, absorptive, interpretative, strategic, excitement)	Joint agenda setting Mobilising actors in a process of dialogue Dealing with distances Bricolage Open innovation Creating informal and multilevel networks

process of shared leadership in which collective values, feelings, trust, commitment and energy form the basis for mobilising private and public actors around a joint agenda. The empirical analysis shows that leaders help to raise awareness in the region and play a sense-making role, telling their stories and trying to align people around issues. The question is to what extent the leaders introduce new sustainable regional agendas.

The scope of the leaders' visions and their actions and ambitions differ from case to case. Depending on the type of coalition, the agenda may be more business-oriented (for example, the New Mixed Business and the Sjalon business partnership) or region-oriented (Het Groene Woud, Northern Frisian Woods). The concept of sustainability varies, from a focus on environmental aspects (New Mixed Business, Northern Frisian Woods) to a broader perspective, including ecological, sociocultural and economic aspects (Het Groene Woud). Sociocultural aspects are underestimated in most cases. Furthermore, the inner motives of the respondents and what drives them to pursue their own particular vision of sustainability differ considerably.

Concepts like the integral care community, rural-urban contracts, the Landzijde model and regional branding are strongly advocated by the leaders of change, but not all are equally good at making effective use of skills such as network power and the ability to seduce others. The leaders of change contribute to the framing of new perspectives. An example is the message in Het Groene Woud – 'to put the region on the map' – which was adopted by the province. Another example is the message in the Northern Frisian Woods project – regional development 'steered by farmers' – which resulted in farmers being given a crucial role in developing and implementing the plans for the National Landscape.

Research on vital coalitions in a local, urban context shows that a 'sense of urgency' is an important aspect of vital coalitions (Tops, 2007). Is this aspect also important in the rural, regional context? As there were no crises, disasters or critical events behind the projects investigated in this study, there was not always a clear sense of urgency. However, we found that a shared 'sense of direction' can stimulate co-operation if private goals are linked effectively to governmental agendas and timing. An example is the Overdiepse polder, in which private plans for water storage fitted in very well with the government's plans to create more 'room for the rivers'. This occurred at a time when the problems of climate change and rising water levels were already high on the political agenda. The success of the leaders of change depends of the extent to which they can create or contribute to a joint sense of direction in their region.

The Dutch change leaders in the cases described in this book act mainly as visionaries and network shuttles, and sometimes as battering rams, in pursuit of their goals. Competences and skills like network capabilities and absorptive, interpretative and socialisation capabilities are used to mobilise people and create teamwork. The importance of using the ability to excite – to generate and capitalise on creative tensions – seems to be underestimated, but is valuable for 'keeping the energy alive'.

Change leaders have a proactive, flexible attitude, are creative, and link their private goals to future perspectives for their region. The qualities they need include patience to maintain momentum in the often lengthy processes, as well as open-mindedness, self-will, self-reflection and perseverance. Motivating people is what they do best, by telling their story over and over again, visioning between visions and contributing to the framing of new issues and agendas. However, they face the risk of exhaustion, running too far ahead of the pack, and getting bogged down in discussions about rules and structures.

The I dimension of leadership – inner motivation, values and passion – is a crucial factor in development processes, but is still underestimated. The initiators in the eight Dutch cases have a clear inner drive, although their motivations vary between a feeling of responsibility, commitment to the region, and religious or spiritual beliefs.

We distinguished three important success factors for creating vital coalitions:
1. Establishing a solid business community of entrepreneurs with a common goal and agenda is a good starting point for forming coalitions with other actors. It is import to maintain the autonomy of individual group members.
2. Backing from government authorities often turns out to be crucial. Some respondents argue that to avoid fragmentation within the provincial organisation, government support should take place from outside the provincial government departments, for example by establishing a development agency with the capacity and powers to implement regional programmes.
3. Reflection, regular communication and the use of formal and informal rules on how to work together are important. Some initiatives organise reflection meetings on a regular basis to share feelings and expectations, explain the roles to be played by those involved and build trust and commitment.

When trying to introduce alternative pathways to the future, leaders of change experience regime power as a glass ceiling of institutions, rules and dominant ideas. In networks they pursue different strategies to remove 'frozen shapes' and lock-ins:
- raising awareness, framing and visioning between visions, telling their story over and over again;
- functioning as boundary spanners and hybrid actors, balancing between their own innovative network and the existing regime;
- open innovation, using knowledge outside their network;
- creating informal and multilevel networks and using network power to mobilise people around a joint agenda and influence decision making, and leapfrog strategies in co-operating with governments.

The results from these case studies show the contours of what we have defined as a route towards a more value-oriented form of shared leadership in regional development. However, further research in different institutional contexts (in other countries) is needed to explore this line of thinking.

References

Bang, H.P., 2004. Culture governance. Governing Self-Reflexive Modernity. Public Administration 82(1): 157-190.
Bauman, Z., 2005. The Liquid Life. Polity Press, Cambridge, UK.
Beckers, T., 2008. Het waren toch mensen. In J. Van der Meer, W. Van Winden and L. Van den Berg (eds.), Stille krachten; 25 jaar sociaal-economische ontwikkeling regio Eindhoven. Lecturis BV, Eindhoven, the Netherlands.
Bragg, M., 1996. Reinventing influence. How to Get Things Done in a World Without Authority. Pitman Publishing, London, UK.
Buchanan, D. and D. Boddy, 1992. The Expertise of the Change Agent: Public performance and backstage activity. Prentice Hall, London, UK.
Burns, J.M., 1978. Leadership. Harper and Row, New York, NY, USA.
Chesbrough, H., 2003. Open innovation, the New Imperative for Creating and Profiting from Technology. Harvard Business School Press, Boston, MA, USA.
Cooperrider, D.L., D. Whitney and J.M. Stavros, 2008. Appreciative Inquiry Handbook For Leaders of Change. Crown Custom Publishing, Brunswick, OH, USA.
Davies, J.S., 2002. Urban regime Theory; A Normative-Empirical Critique. Journal of Urban Affairs 24(1): 1-17.
Derix, G., 2003. Hof der Onthaasting. Werkboek, Orbis, Sittard, the Netherlands.
Douglas, M.T. (1986). How institutions think. Syracuse University Press, Syracuse, NY, USA.
Dowding, K., 2001. Explaining Urban Regimes. International Journal of Urban and Regional Research 25(1): 7-19.
Dryzek J.S., 2005. Politics of the Earth: Environmental discourses. Oxford University Press, Oxford, UK.
Duffhues, T., 2009. Waarde(n)volle gebiedsontwikkeling: werken met 'hart en ziel'. In: I. Horlings, G. Remmers and T. Duffues (eds.). Bezieling: de X-factor in gebiedsontwikkeling. Salsedo, Breda, the Netherlands, pp. 44-61.
Elzen, B., C. Leeuwis and B. Van Mierlo, 2008. Anchorage of innovations: Using the greenhouse effect to save energy. Paper presented at the conference: Transitions towards sustainable agriculture, food chains and peri-urban areas. Wageningen University and Research Centre in Wageningen, the Netherlands.
Frouws, J., 1998. The contested redefinition of the countryside. An analysis of rural discourses in the Netherlands. Sociologia Ruralis 38(1): 54-68.
Goodwin, M. and J. Painter, 1997. Concrete research, urban regimes and regulation theory. In: M. Lauria (ed.), Reconstructing Regime Theory; regulating urban politics in a global economy. Sage, London, New Delhi, pp. 13-29.
Hart, P. 't, 1999. Hervormend leiderschap in het openbaar bestuur. Inaugural lecture, Leiden University, Leiden, the Netherlands.
Heifetz, R.A., 2003. Leadership without easy answers. Belknap Press of Harvard University Press, Cambridge, MA, USA.
Horlings, I. and M. Van Mansfeld, 2006. Back to BSIK? Procesbeschrijving Integrated Project Green Valley. Telos and TransForum, Tilburg, the Netherlands.
Horlings, I., A.J. Bijsterveld and J. Janssen, 2006. Het Groene Woud is vele landschappen. In: Zoeken naar nieuwe wegen, Telos and TransForum, Tilburg, the Netherlands, pp. 63-75.
Horlings, I. and W.M.F. Haarmann, 2007. The soft stuff is the hard stuff; vital coalitions in rural-urban regions. Paper for the conference of the European Society of Rural Sociology, Wageningen, the Netherlands.

Horlings, I. and W.M.F. Haarmann, 2008. Botsende belangen en vitale verbindingen. Position paper over het Integrated Project Nieuwe Markten en Vitale coalities Heuvelland, In: TransForum, Nieuwe Markten en vitale coalities Heuvelland Zuid-Limburg, innovatief praktijkproject, Workingpaper no. 8.: TransForum, Zoetermeer, the Netherlands, pp. 95-174.

Horlings, I., 2008. Strategieën en rollen van private voortrekkers in duurzame gebiedsontwikkeling. In: B. Dankbaar, B. Kessener and J. Boonstra (eds.), Doelgericht veranderen in bedrijf en samenleving in de 21ste eeuw, Kluwer, Deventer, the Netherlands, pp. 43-60.

Horlings, I. and W. Haarmann, 2009. Makelen en schakelen tussen schillen en schalen; De leerervaringen van het TransForum project Ons Zuid-Limburgs land, Telos, Tilburg, the Netherlands.

Haarmann, W., I. Horlings and G. Derix, 2009. Please in my backyard, Parc Hoogveld en de opkomst van de integrated care community, TransForum, Zoetermeer, the Netherlands.

Horlings, I., 2009. The role of leaders of change in the transition towards sustainability in rural-urban regions in The Netherlands. Paper presented at the 1st European Conference on Sustainability Transitions: Dynamics & Governance of Transitions to Sustainability. Felix Meritis, Amsterdam, the Netherlands.

Horlings, I. and F. Padt, 2009. Leadership in rural-urban networks. Paper for the ISSRM conference, Vienna, Autrsia.

Horlings, I., P. Tops, J. Van Ostaaijen and E. Cornelissen, 2009a. Vital coalitions. In: H. van Latesteijn and H. Mommaas (eds.), The Organisation of Innovation and Transition. Workingpaper no. 2, TransForum, Zoetermeer, the Netherlands, pp. 3-44.

Horlings, I., G. Remmers and T. Duffhues, 2009b. Bezieling; de X-factor in gebiedsontwikkeling. Salsedo: Breda.

Huber, G.P. and R.L. Daft, 1987. The information environments of organizations. In: F.M. Jablin, I.L. Putnam, K.H. Roberts and L.W. Porter (eds.), *Handbook of Organizational Communication: An Interdisciplinary Perspective*. Sage, Newbury Park, CA, USA, pp. 130-164.

Jaworksi, J., 1998. The Inner Path of Leadership. Berret-Koehler Publishers, San Francisco, CA, USA.

Jones, R. and G. Noble, 2008. Managing the Implementation of Public-Private Partnerships. Public Money & Management, 28(2): 109-114.

Kanter, R.M., 1999. The Enduring Skills of Change Leaders. Leader to Leader No. 13: 15-22. Available at http://www.pfdf.org/leaderbooks/l2l/summer99/kanter.html.

Kickert, W.J.M., E.H. Klijn and J.F.M. Koppenjan, 1997. Managing complex networks: strategies for the Public Sector. Sage, London, UK.

Klein, E. and G.R. Teisman, 2003. Institutional and Strategic Barriers to Public-Private Partnership: an Analysis of Dutch cases. Public Money and Management 23(3): 137-146.

Lauria, M., 1997. Reconstructing Urban Regime Theory. Regulating Urban Politics in a Global Economy. Sage Publications, Thousand Oaks, CA, USA.

Marsden, T., 1999. Rural futures: the consumption countryside and its regulation. Sociologia Ruralis 39(4), 501-520.

Millennium Ecosystem Assessment, 2005. Ecosystems and Human Well-being: Synthesis. Island Press, Washington, DC, USA.

Noble, G. and R. Jones, 2006. The role of boundary-spanning managers in the Establishment of public-private partnerships. Public Administration 84(4): 891-917.

OECD (Organisation for Economic Cooperation and Development), 2006. The new rural paradigm: policies and governance. Paris, France.

Padt, F.J.G., 2006. Regional environmental planning in the Netherlands: un unstable settlement of policy arrangements. In: B. Arts and P. Leroy (eds.), Institutional Dynamics in Environmental Governance. Heidelberg: Springer, pp. 203-223.

Padt, F.J.G., 2007. Green Planning. An institutional analysis of regional environmental planning in the Netherlands. Dissertation. Eburon, Delft, the Netherlands.
Rowley, R. and J. Roevens, 1999. Organize with Chaos. Management Books 2000 Ltd, Gloucestershire, UK.
Scharmer, O., 2005. Theory U: Learning from the Future As It Emerges. Berrett-Koehler Publishers, San Francisco, CA, USA.
Schön, D.A., 1973. Beyond the Stable State. Public and private learning in a changing society. Penguin, Harmondsworth, London, UK.
Schön, D.A., 1983. *The Reflective Practitioner. How professionals think in action.* Temple Smith, London, UK.
Scott, W.R., 2001. Institutions and organisations. Sage Publications, Thousand Oaks, CA, USA.
Sotarauta, M., 2002. Leadership, Power and Influence in Regional Development. A Tentative Typology of Leaders and their Ways of Influencing. In: M. Sotarauta and H. Bruun Nordic (eds.), Perspectives on Process-Based Regional Development Policy. Stockholm: Nordregio, pp. 182-207.
Sotarauta, M., 2005. Shared Leadership and Dynamic Capabilities in Regional Development. In: I. Sagan and H. Halkier (eds.), Regionalism Contested; Institution, Society, Governance. Ashgate Publishing, Aldershot, UK, pp. 53-72.
Sotarauta, M., 2006. Where Have all the people gone? Leadership in the Fields of Regional Development. Sente-Working papers 9/2006. University of Tampere Research Unit For Urban and Regional Development Studies. Tampere, Finland.
Stoker, G., 1995. Regime theory and Urban Politics. In: D. Judge, G. Stoker and H. Wolman (eds.), Theories of Urban Politics. Sage Publications, London, UK, pp. 54-71.
Stone, C.N., 1989. Regime Politics: Governing Atlanta, 1946-1988. Kansas University Press, Lawrence KS, USA.
Stone, C.N. 2002. Urban Regimes and Problems of Local Democracy, paper prepared for Workshop 6, Institutional Innovations in Local Democracy, ECPR joint sessions, Turin, Italy.
Tops, P., 2007. Regime verandering in Rotterdam; hoe een stadsbestuur zich opnieuw uitvond. Uitgeverij Atlas, Amsterdam, the Netherlands.
Van de Poel, I., 2000. On the Role of Outsiders in Technical Development. Technology Analysis & Strategic Management 12(3): 383-397.
Van Mansfeld, M. and H. Van der Stoep, 2007. Procesverslag Heuvelland. Wageningen University and Research Centre, Wageningen, the Netherlands.
Weick, 1995. Sensemaking in Organizations. Sage, Thousand Oaks, CA, USA.
Wielinga, H.E., 2005. Het assisteren van kennisnetwerken, Netwerken, verbindingen en interventiestrategieën. Onderdelen A en B van het project Kennis over Netwerken - Projectcode 30303. Landbouw Economisch Instituut, Den Haag, the Netherlands.
Wielinga, H.E., 2001. Netwerken als Levend Weefsel. Een studie naar kennis, leiderschap, en de rol van de overheid in de Nederlandse landbouw sinds 1945. Dissertation nr. 2963. Wageningen University and Research Centre, Wageningen, the Netherlands.
Wilber, K., 2000. Integral Psychology; Consciousness, Spirit, Psychology, Therapy. Shambhala Publications, Boston, MA, USA.
Wiskerke, J.S.C., 2007. Robuuste regio's: dynamiek, samenhang en diversiteit in het metropolitane landschap. Inaugural lecture, Wageningen University, Wageningen, the Netherlands.
VROM-raad, 2004. Meerwerk; advies over de landbouw en het landelijk gebied in ruimtelijk perspectief. Advies 042, VROM, Den Haag, the Netherlands.
Zimmerer, K.S. (ed.), 2006. Globalization and new geographies of conservation. Chicago University Press, Chicago, IL, USA.

Chapter 6
New Markets in Heuvelland: coalition building and agenda setting

Hetty van der Stoep and Noelle Aarts

In this chapter we reflect on the New Markets initiatives in the rural-urban region of Heuvelland in the Netherlands. More specifically, we analyse whether and how these mainly private initiatives developed an agenda-setting capacity for sustainable regional development. Drawing on the findings about agenda-setting capacities, including the obstacles encountered by the initiatives, we make recommendations for the organisation of vital coalitions.

6.1 Introduction

In Chapters 2 and 3 vital coalitions were described as coalitions with a capacity to achieve outcomes, a capacity to act. How coalitions develop a capacity to act and how they may challenge policy regimes is the focus of the empirical chapters. In this chapter we address the ability of private-public initiatives to construct an appealing agenda that mobilises support. The New Markets in Heuvelland project provided information for an in-depth case study about initiatives that try to develop capacity to act to achieve their goals and thereby confront existing policies for regional development. New Markets in Heuvelland is an initiative by the Limburg provincial government to bring together a group of entrepreneurs from different economic sectors with the aim of boosting the rural economy in the region, while safeguarding or strengthening environmental quality. For some, the project was about an innovative approach to area development, starting from economic activities instead of spatial visions. The research question for this case study is:

> How did the New Markets initiatives develop agenda-setting capacity for sustainable regional development in Heuvelland?

6.1.1 Analytical framework

In theory, vital coalitions bring forward innovative ideas about sustainable regional development and thereby challenge existing policy agendas for regional development or regional regimes (see Chapter 1). Our aim, therefore, was to study the agenda-setting mechanisms and strategies adopted by the New Markets initiatives. Our approach, as described below, was inspired by policy agenda-setting theory and insights from framing theory.

Agenda setting is generally defined as an ongoing competition between issue proponents for the attention of media professionals, the public and policy elites (Dearing and Rogers, 1996). At stake is the relative attention given by the media, the public and policy makers to some issues and not to others. According to Kingdon (2003) policy agendas can be changed when the relatively independent streams of problems, policy solutions and politics are connected to each other. Connections between those relatively separate worlds are made by 'policy entrepreneurs', who are willing to invest their resources in return for future policies they favour or for material, purposive or solidary benefits. These policy entrepreneurs develop strategies to push proposals and issues onto the policy agenda and to mobilise support. Moreover, they are able to make the necessary couplings between the streams of problems, policy solutions and politics when windows of opportunity emerge. Windows of opportunity are 'trigger events' concerning problems that require immediate attention, such as calamities, a swing in the national mood or shifts in the political sphere, such as elections.

The main strategies adopted by policy entrepreneurs or issue proponents to influence the policy agenda are (Kingdon, 2003; Mintrom, 1997): (1) identifying problems; (2) choosing issues to be pushed or not pushed, depending on the target audience; (3) networking in policy circles to learn about the 'world views' of members of the policy community and to develop arguments in favour of the proposal; and (4) building coalitions to get supporters for the idea. Summarising, we need to consider two important agenda-setting elements: the way issues are construed and the way coalitions and relations between actors are built to get support for the proposed issues.

Issue construction, which starts with problem identification, is probably the most powerful agenda setter (Jones and Baumgartner, 2005; Kingdon, 2003). 'The chances of a given proposal or subject rising on an agenda are markedly enhanced if it is connected to an important problem' (Kingdon, 2003: 198). Therefore, an important agenda-setting strategy is getting people to frame a 'situation' as a 'problem' by adding colour, in other words, by using a frame that highlights certain aspects of the situation (Jones and Baumgartner, 2005). Framing of issues, stressing one perspective and ignoring others, is considered to be a powerful force in agenda-setting processes (Dearing and Rogers, 1996; McCombs and Shaw, 1993; Scheufele and Tewksbury, 2007; Weaver, 2007). A generally accepted definition of framing is 'to select some aspects of a perceived reality and make them more salient in a communicating text, in such a way as to promote a particular problem definition, causal interpretation, moral evaluation, and/or treatment recommendation for the item described' (Entman, 1993: 2). Framing and discourse studies frequently address the construction of storylines as a mobilising force that guides actions and brings people together in coalitions (Hajer, 1995; Rein and Schon, 1996). Storylines or narrative frames are strong because they connect themes to each other and incorporate and adapt to changing events. They can re-order understandings and may even lead to political change (Rein and Schon, 1996). This implies that agency, or in this case the efforts of private initiatives to achieve their goals, should be aimed at constructing an appropriate storyline that mobilises supporters with access to the necessary resources.

Apart from the framing of issues, framing is also relevant for coalition building, the second element in our study of agenda setting. Besides substantive issues, relationships and interaction situations and events can also be framed (Dewulf et al., 2009; Entman, 2003). Therefore, framing can also apply to the process of coalition building and relations between actors. For example, who should be 'in' and 'out', and which kinds of relations will or will not benefit coalitions that are being built around issues. We are particularly interested in the effect of relation framing on coalitions that are being built or deconstructed, and ultimately why and how this affects agenda setting.

To conclude, we have identified two factors that we need to consider in our study of how the New Markets initiatives developed agenda-setting capacity. First, agenda setting involves issue framing, which involves problem identification as well as selecting a particular perspective from which to promote the issue. Second, agenda setting involves coalition building to get supporters for the issue and improve the chances for pushing the issue onto the policy agenda. We suggested that coalitions are built around powerful and appealing storylines which somehow address the collective goal of the coalition. However, we also need to consider the relations between people when studying how coalitions are constructed. Framing provides a way to study both coalition building and issue construction in agenda-setting processes. The analytical framework is illustrated in Figure 6.1.

6.1.2 Methodology

This analytical framework results in a methodological focus on framing as an agenda-setting mechanism. How did issues develop from the interactions that took place in the New Markets in Heuvelland project? How did key stakeholders talk about the goals and how to go about attaining them? What effect did this have on the coalitions that were built and the issues that were being promoted? We used framing analysis to identify frames about issues that were considered important for the construction of the agenda, and frames about the procedural conditions to achieve the goals. We attended meetings of the New Markets initiatives, which were recorded in reports. We studied policy and other documents to see how issues developed on the regional policy agenda. And finally, we interviewed 17 key players involved in and relevant to the New Markets initiatives. Interviews with participants in the New Markets process (project managers, entrepreneurs and a representative from the Provincial Council's economic affairs department) generated data about how issues developed during the process, how this related to the goals of the participating parties, and frames about the construction of the coalitions during the New Markets process. Interviews with executive councillors,

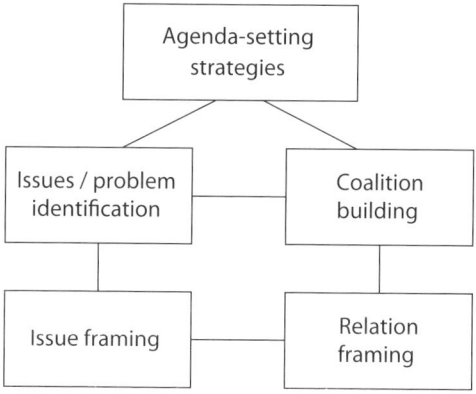

Figure 6.1. Analytical framework for the New Markets study.

provincial government officers and a representative from a nature conservation organisation generated data about dominant policy frames on regional development and about whether and how the New Markets initiatives affected the ideas and policies of regional and local authorities and other decision-making bodies.

6.2 Issue development

In this section, we describe and explain how the agenda of the New Markets initiatives changed, as well as the coalitions in and related to the New Markets process. Moreover, we explain why the initiatives were or were not successful in setting or influencing the policy agenda for regional development.

6.2.1 Start of the New Markets in Heuvelland initiative

The New Markets in Heuvelland project was launched in 2004 in response to economic problems in the rural-urban region of Heuvelland (in the southern tip of the province of Limburg). The region covers about 30,000 hectares and lies between Belgium and Germany. It has a population of 600,000 inhabitants and after the *Randstad* in the western part of the Netherlands is the second most urbanised area in the Netherlands. The three cities of Maastricht, Sittard/Geleen and Heerlen form an urban crescent around the northern half of the rural landscape, which is commonly called Heuvelland (see Figure 6.2).

Urban pressures in the area are causing increasing fragmentation of the landscape and worsening the conditions for agricultural activities. Another problem in the region is the shrinking population, leading to vacant properties, including valuable historic farmsteads and cloisters. The provincial government is tackling these problems through a long-term programme for rural development. In 2005, Heuvelland was designated a National Landscape by central government, which provides an extra stimulus to protect the landscape qualities of Heuvelland. The province has adopted a Landscape Vision as a framework for subsidised investments in the Heuvelland National Landscape.

Another pillar in the provincial policy for regional development is economic development. The closure of the coal mines during the 1960s and 1970s and the resulting socioeconomic problems triggered policies for economic development and employment creation that are still being pursued. Stimulation programmes were designed to favour certain economic activities, first the chemical industry and services, and later medical technology, transport and tourism. In 2005, the province adopted its Acceleration Agenda (*Versnellingsagenda*) to put its economic difficulties in the past and develop an economic advantage over other Dutch and European regions. The 'new economy' is the latest buzz word in the region, reflecting a desire to create opportunities for the development of hi-tech industries, for example in the field of medical technology.

Besides technology development, tourism is also an important economic sector in the region. In the late 19th century, South Limburg was the first region in the Netherlands to attract tourists and by the beginning of the 1990s tourism accounted for 18% of the region's economy,

6. New Markets in Heuvelland

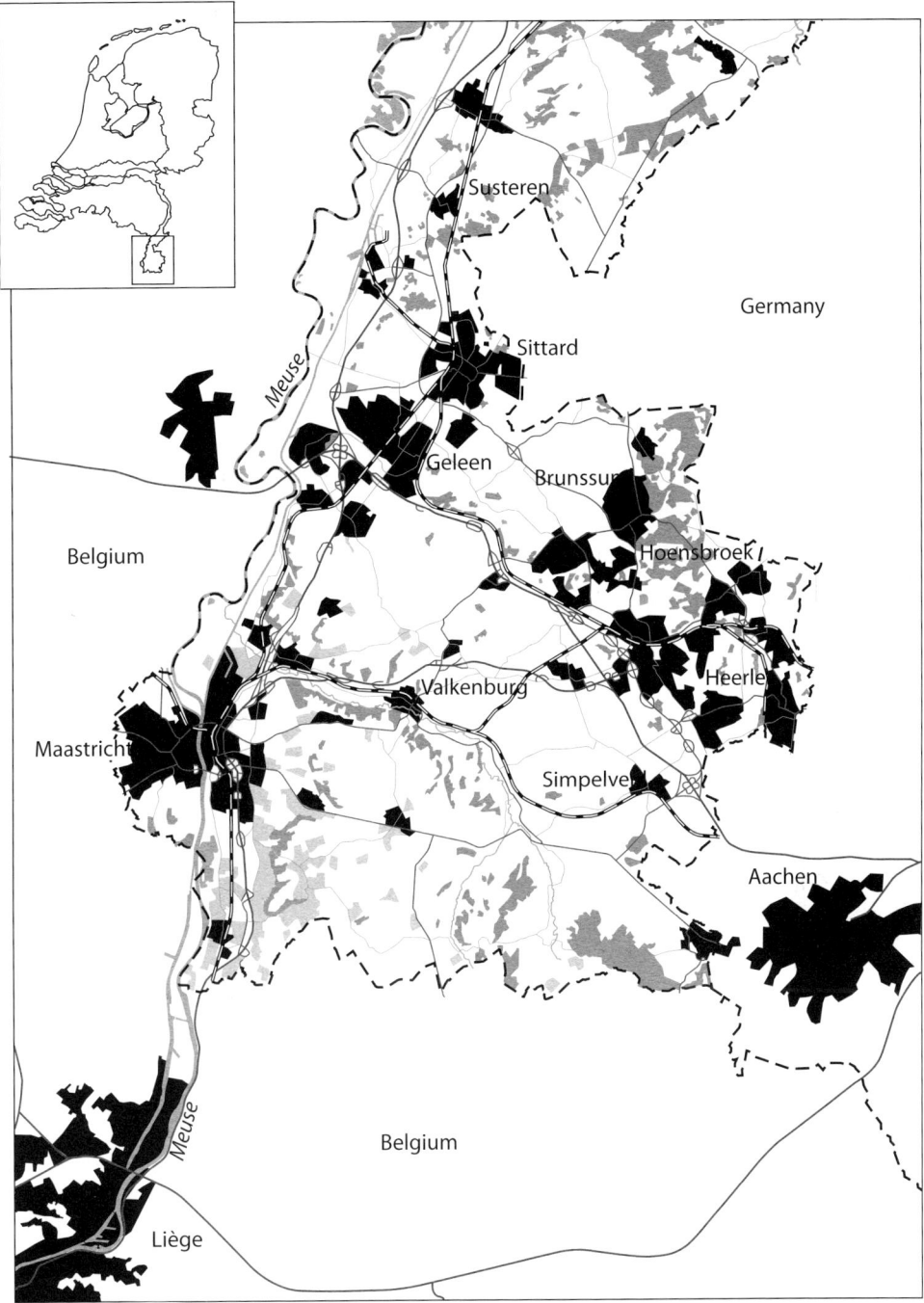

Figure 6.2. Map of Heuvelland, South Limburg.

but this declined to 10% in 2004 (ZKA *et al.*, 2005). The decline was said to be caused by structural problems in the tourism sector, such as increasing competition, the increasing preference of tourists for foreign and more exotic destinations, the small scale of the tourist offer in South Limburg and the outdated quality of the offer compared with other regions (Anonymous, 2005; Mommaas and Janssen, 2008). Moreover, experts and public officials argued that tourism was being threatened by a deterioration in the rural landscape due to the increasing scale of agricultural production and urbanisation, which makes the landscape less attractive to tourists. The provincial government argued it was necessary to do something to improve the tourism economy that would also preserve the quality and identity of the region. This resulted in the New Markets in Heuvelland initiative.

The New Markets project was launched in 2004 by Limburg Provincial Council to deal with the related problems of the tourism economy, the changes in agriculture and the resulting changes in the landscape. The province delegated the task of devising new incentives for tourism to the Limburg Investment and Development Agency (LIOF, *Limburgse ontwikkelings- en investeringsmaatschappij*). Although LIOF was specialised in other economic sectors, like technology development, and did not have much know-how in tourism and the primary sector, the province had no other choice because the Limburg Tourism Bureau had just been closed. LIOF invited a number of experts to advise on the matter. They came up with the New Markets in Heuvelland programme to connect tourism to the landscape values of South Limburg and to businesses with the financial resources to sustain and further develop the sector. LIOF and the panel of experts decided to offer private entrepreneurs as much opportunity as possible to develop their own ideas on the economic regeneration of Heuvelland. They were afraid that the process might become bogged down in discussions between provincial government officers and executive councillors, which is why the provincial government did not participate actively in the planning process (Horlings and Haarmann, 2008; Mommaas and Janssen, 2008).

The experts selected themes that could serve as 'new markets', or new product-market combinations, which were then presented in round-table meetings with private entrepreneurs and investors who had already proved to be innovative, active in the region, financially strong and willing to invest. The goal of the meetings was to test how attractive the proposed new markets were to the entrepreneurs and to convince them to develop the ideas into business plans. The meetings resulted in the selection of five new markets (see the Box 6.1), each with an entrepreneur who was willing to invest resources to further develop and implement the idea. This led to some projects that were successfully implemented, such as a hotel package for surgical patients and a regional food market in the city of Maastricht.

After the initial energy that flowed from the identification of the new markets, action by the entrepreneurs tailed off. The activities that were undertaken tended to be individual projects of the entrepreneurs that did not generate spin-offs, nor did they scale up to create new economic perspectives for the region involving collaboration between entrepreneurs from different sectors (Horlings and Haarmann, 2008; Van Mansfeld and Van der Stoep, 2008). At this point, the province thought it would be a good idea to find other ways of injecting new life into the New Markets project. The aim this time was to develop a sustainable business community and a system for financing the New Markets in Heuvelland projects (Horlings and Haarmann, 2008). With the assistance of the experts involved in the initial stage of

> **Box 6.1. The five new markets.**
>
> - *Healing Hills*: linking the natural and cultural values of the rural area (hills, meadows, woods, convents) and healthcare, specifically by providing hotel accommodation for recovering surgical patients.
> - *Rich Tastes*: linking the production of food on farms in Heuvelland to distribution systems that service the surrounding cities. Regional agricultural products would be distributed in a regional chain of quality restaurants, gastronomy courses, etc.
> - *Wellness in Luxury*: investing in preventive care activities by healthcare insurance companies (e.g. health resort stays, therapeutic baths, training courses).
> - *Glorious Life*: matching the extra demand for senior housing and related support services resulting from the demographic trend towards an ageing population in the region to the large number of empty farms and convents. This warrants financial support for the preservation of the historical values of those buildings.
> - *Linked Fields*: relating multimedia services to geographical information about the location of tourist sites, events, production of regional food, healthcare services, etc.

New Markets, the province established a partnership with Transforum, a state-financed organisation that provided subsidies and know-how for promising projects that support the transition towards sustainable development in rural regions. The alliance with Transforum not only brought in money, but also the participation of researchers from Alterra research institute at Wageningen UR, who would contribute knowledge about rural transition processes and organise multistakeholder co-operation.

6.2.2 A new coalition partner and a new agenda

The coalition with Transforum and Alterra resulted in a shift in emphasis in the goals and method of the New Markets project. The project was initially launched to tackle the problem of declining tourism and the related threat to the quality of the landscape. When Transforum and Alterra were brought in, attention shifted to conserving and strengthening the unique selling points of the region – the nature and landscape qualities – as a condition for improving the regional economic structure. The question was how the new markets could generate financial support to maintain the landscape. The process was managed by a project team consisting of a representative from Alterra, an independent consultancy representing the entrepreneurs and a representative from the provincial government's economic affairs department. In view of the goal of securing and strengthening landscape qualities, Alterra wanted to bring other stakeholders into the process: the provincial spatial planning and rural development departments, and representatives from nature and landscape organisations and farmers' organisations. This shift in problem definition, goals and process are illustrated in Table 6.1.

The shifts in emphasis in the problem definition and the project goals and process were not explicitly discussed by Alterra/Transforum and the entrepreneurs and representatives from the provincial government's economic affairs department. However, during the project the involvement of stakeholders from the field of landscape management was continuously contested within the project team. The other project team members were afraid that a broadened agenda and inclusion of more stakeholders would make the process too complex

Table 6.1. Problem definition and goals of the New Markets in Heuvelland project in 2005 and in 2006 (ZKA et al., 2005; Van Mansfeld and Van der Stoep, 2008).

	Initial problem definition and goals of New Markets in Heuvelland 2004	Problem definition and goals after the introduction of Transforum 2006
Problem definition	Tourism is declining because of the changing preferences of tourists, accessibility problems, decreasing capacity, and the small-scale and fragmented tourist offer. The sector is not able to anticipate to the market. Tourism businesses are closing, with a change of use to other functions (housing, employment) or are growing in size, leading to a loss of environmental quality.	Problem definition Economic drivers and traditional managers of landscape qualities are disappearing; agriculture and tourism are declining. Urbanisation puts pressure on the conservation of landscape qualities in the open area between the cities. These related and complex problems threaten the 'unique selling points' of Heuvelland, which are the starting point for improving the economic structure of the region.
Goals	Improving the structure of the tourism economy of Heuvelland in combination with conserving or strengthening environmental quality and cultural identity.	Safeguarding and strengthening of the environmental qualities of Heuvelland by achieving a new investment dynamic through a vital coalition of economic drivers.
Ideas about the intended 'vital' coalition	Developing horizontal alliances between tourism and sectors with dynamic economic activity, new initiatives and investment capital, and which are able to provide an economic basis for both the tourism sector and environmental quality.	Facilitation and realisation of new higher-level coalitions within the region, supported by all stakeholders, who will be jointly responsible for the conservation, maintenance and development of the unique selling points of the landscape.

and slow for the entrepreneurs, who might then lose interest because their investments of time and money would be unlikely to generate concrete outcomes in the short term.

> *I wonder whether I understood correctly what the line of thinking of New Markets was. I did write a proposal with the representative from the provincial economic affairs department…. However, the approach taken by this department was: begin with profit, which will bring the others in. The problem definition of our project was not recognised by the entrepreneurs. Most entrepreneurs only think about earning money.*
>
> (Project team member from Alterra)

Disagreement in the project team about the organisation of the process hampered the organisation of workshops with entrepreneurs and other stakeholders. During the first phase of New Markets in Heuvelland, attempts to involve representatives from the spatial planning and rural development departments failed. Civil servants and government officials from the economic affairs department, who were responsible for New Markets, and from the spatial planning and rural development departments showed little interest in co-operation. As the conflicts over the process architecture started to dominate discussion in the workshops, little energy was invested in defining the ambitions of the project for economic and spatial regional development. As a result, the agenda of the New Markets process was not clear to the participants or the other stakeholders. We will come back to this topic in Section 6.3.

6.2.3 'Healthy Living in South Limburg' the winning theme

By the end of this phase of the New Markets process (November 2006), the process had produced few outcomes and this frustrated all the stakeholders. The executive councillor for economic affairs decided to intervene, as he would soon have to account for the investments made by the province in the process. The urgency to do something was heightened because the term of office of the Provincial Executive was nearing its end. The executive councillor for economic affairs pinned his hopes on one of the participating entrepreneurs, the director of the healthcare concern Orbis. In the last workshop of the New Markets process, at request of the economic affairs representative in the project team, the Orbis director presented his concept for Healthy Living in South Limburg. This had not been discussed with the rest of the project team. Not only was the Orbis director given the opportunity to present his idea for a business plan, but he was also presented as the new project leader for the next phase of the New Markets process.

The problem definition and goals of the Healthy Living in South Limburg initiative were more narrowly defined than the goals drawn up by Alterra for the New Markets project. The Healthy Living in South Limburg proposal did involve new product-market combinations, but they were all health-related and scarcely addressed how these new markets could contribute to the management of landscape qualities in Heuvelland. Nevertheless, this initiative was prioritised by the Provincial Executive, without consulting the project team or the wider circle of stakeholders. Why did this happen? We argue that developments in the policy and political context of the provincial government were decisive in prioritising the Healthy Living in South Limburg initiative.

In his presentation, the Orbis director based his proposal for Healthy Living in South Limburg on the problem of the declining market for healthcare in the region. An increasing number of people were turning to facilities in Belgium and Germany, which was having adverse effects on the healthcare services in South Limburg. He believed that the market for facilities in South Limburg could be improved by focusing on the growing number of older and chronically sick people in the country. His idea was to combine healthcare services with tourist and housing accommodation. The combination with tourism drew on the existing Healing Hills theme in the New Markets in Heuvelland project (rehabilitation and convalescence in hotel accommodation in Heuvelland). Moreover, the Orbis director saw opportunities to combine healthcare with new housing facilities for older people (the 'new market' Glorious Life), linking the problem of the healthcare market to the economic problem of the shrinking population in Heuvelland. In his view, the migration of young people to the west of the Netherlands could not be prevented. Instead, efforts should be focused on the new market of older people, not only people from Heuvelland, but also from other regions. This could be done by combining new housing facilities with customised health services. Moreover, new housing could be combined with the conservation of cultural heritage (cloisters, farmsteads), and thereby contribute to the conservation of landscape qualities.

The idea of focusing on the opportunities presented by the ageing population, investing in the older consumer market, reflected the ambitions of the provincial government to establish an economy based on older consumers: the 'silver economy'. The Healthy Living in South Limburg initiative fitted in well with the goals of the provincial Acceleration Agenda and the Provincial Executive's policy programme for 2003-2007. The Acceleration Agenda, which consists of core themes for future economic development, including Health, Care & Cure, was considered the dominant steering mechanism for the economic development of the region.

> *In 2012 South Limburg will have developed much new added value in the so-called Health, Care & Cure market. By providing innovative products and services Limburg will help the ageing population of North West Europe to become and stay healthy.... The active policy for the older consumer market will deliver benefits and the business community in Limburg will lead the development of new products and services for this target group. Limburg has more than enough lifetime homes with comfort, wellness and care arrangements in or close to the home. Tourism will also play its part. In 2012 older tourists will prefer Limburg for accommodation and consumption because the tourism products and services are also accessible to the less active among them.*
>
> (Taskforce Versnellingsagenda 2005: 11-15)

Even before the Acceleration Agenda of 2005, the Provincial Executive had been focusing on the opportunities presented by the ageing population for the regional economy, as evidenced by the Provincial Executive's policy programme for 2003-2007.

> *Limburgers are encouraged to be enterprising and a self-confident in seizing the opportunities for a technological top region (industry and agribusiness), for a stimulus to the quality of tourism offer and for a 'silver economy', aimed both at*

retaining elderly people for the labour market and product innovation for older consumers (ICT, healthcare, leisure, service).
(Taskforce Versnellingsagenda 2005: 21)

By the end of the Provincial Executive's term of office, the executive councillor for economic affairs had to deliver results for both the New Markets project and the economic policy presented in the Provincial Executive's policy programme and the Acceleration Agenda. Prioritising the Healthy Living in South Limburg initiative was most profitable for him in the context of that moment.

6.2.4 Healthy living dropped from the agenda

Only a couple of months after the prioritisation of the Healthy Living in South Limburg initiative, support from the provincial government came to an end. The executive councillor for economic affairs retired after the provincial elections that took place in the spring of 2007. The new Provincial Executive shifted the focus of economic development policy, considering the ageing population not to be an opportunity, but a threat to the development of the desired knowledge economy.

The knowledge economy was already a key issue for the previous Provincial Executive. The ambition was to play a central role in a 'technological top region' which includes neighbouring regions in the Netherlands, Germany and Belgium. As part of this larger region, South Limburg would specialise in life science technologies and creative industries. The new Provincial Executive continued this focus on the knowledge economy and the technological top region. However, demographic trends had become an increasing concern during the past two years, as the flow of talented young people into the labour market had been decreasing. Not only was the population ageing and shrinking more in South Limburg than anywhere else in the Netherlands, but young people were migrating to other parts of the Netherlands to find jobs and good living environments. New businesses and industries in the field of healthcare technologies and creative industries could never develop if there were no talented employees to hire. The Provincial Executive therefore decided to put attracting young people above the provision of services and products for older people. As the dominant economic policy, the knowledge economy pushed the 'silver economy' from the agenda.

As in the previous period, the new Provincial Executive's policy programme for 2007-2011 focused heavily on meeting the challenge presented by the alarming demographic trends: a shrinking and ageing population, and emigration of young people. However, instead of focusing on the ageing population and the older consumer market as a driver of economic development, attention now turned to keeping young people in the region and making them aware of its attractions for employment, housing and living in South Limburg.

Limburg should become a demographically experimental region, because it will be many years before the rest of the Netherlands will be confronted with the same demographic trends. ... We want to appeal to people's talents and particularly to offer a challenging environment to young people
(Provincie Limburg, 2007: 5-6)

The Acceleration Agenda was renewed in 2008. Emphasis was placed on retaining young people in the region as a condition for the development of an innovative and enterprising knowledge economy:

> *The availability of sufficient knowledge workers is indispensable for realising the innovation ambitions in the Acceleration Agenda. However, the demographic trend is working against us.… If we let that take its course, we face a growing quantitative and qualitative shortage of knowledge workers. The influx of young talent from other regions, countries and continents…will not resolve this shortage if we cannot retain this talent in the region. Of course we are pleased to observe that our graduates are global citizens and so well qualified that they can get a job anywhere. But it would be even more pleasing if these global citizens could find their dream job here! Unfortunately, the region is failing on this point. Active programmes are necessary to upgrade current employees…to attract knowledge workers from elsewhere…and keeping talent in the region.*
> (Raad van Advies Versnellingsagenda, 2008: 43)

The three clusters, High Tech Materials, Health, Care & Cure, and Agrofood and Nutrition, were continued, but the emphasis now was on specific projects and implementation by a limited number of institutions and companies. For the Health, Care & Cure cluster this meant that the previously broad scope, which included housing and leisure services for elderly people, was narrowed down to healthcare technology and related businesses.

> *The ageing of the population in Limburg is going faster than in other parts of the country. Ageing results in more demand for healthcare, especially chronic care. Moreover, this care increasingly has to meet the wishes of individuals.… Innovative top healthcare will be combined with developments in areas like biomedical materials, molecular medical science and molecular equipment.*
> (Raad van Advies Versnellingsagenda, 2008: 16-18)

As the Healthy Living in South Limburg initiative had previously been associated with the 'silver economy', it was framed by government representatives as an initiative that would create facilities only for the elderly and the sick and not for young healthy people who needed a 'dynamic environment'. It was therefore not considered appropriate in the light of the new policy emphasis, a case that was vividly supported by the image of a 'rollator landscape':

> *A care landscape alone can have negative effects. The image of stumbling over wheelchairs and rollators is not positive. We have to think outside the theme of healthcare and we should just achieve good living environments.*
> (A former local government official)

> *A municipality in the region is postponing the decision on the building permit for a care farm. They are afraid that it will lead to a rollator landscape.*
> (One of the entrepreneurs, recorded in meeting minutes)

The knowledge economy, and the development of a care landscape have become perceived as two different things. This is utterly inexplicable, because the knowledge economy and a care landscape can complement each other very well. All studies on the knowledge economy affirm the importance of quality of life. A care landscape also refers to quality of life; it is about preventive health care aimed at the well-being of both young and elderly people.

(Chairman of the new Markets in Heuvelland project phase I)

The negative impact a 'care landscape' could have on the new goal of creating a 'dynamic top technological region' was used to justify the removal of the Healthy Living in South Limburg initiative from the economic policy agenda.

Although the province no longer supported the Healthy Living in South Limburg initiative, Transforum still saw potential in further developing this initiative. The withdrawal of the provincial government did not bother Transforum because they felt that in the previous phase the co-operation with the province had caused delay and had not led to any useful discussions. Orbis and Alterra drew up a revised prospectus for the Healthy Living in South Limburg initiative that proposed building Integrated Care Communities. New partners were introduced: an investment bank with interest in rural estates, a property developer and an agricultural estate agency. The provincial government was excluded from this new coalition.

6.2.5 A new attempt to put New Markets on the regional agenda via the local authorities in Heuvelland

In the course of 2006, the local authorities, or municipalities, in the *Heuvelland* region decided to develop their own vision for regional development in response to the provincial agendas for the National Landscape and economic development. They organised themselves into a body called Core Area South Limburg. The New Markets initiative, as a new approach to improving tourism, was also discussed. Core Area South Limburg was co-ordinated by an independent ex-mayor, who was assigned by the provincial government to act as a liaison officer between the province and the rural municipalities. The municipalities agreed to develop a 'grand design' for the region in which new economic activities, as proposed in New Markets, would be a way to implement the proposals in the provincial Landscape Vision for conserving and developing landscape qualities. One of the experts who developed the New Markets in Heuvelland project was invited to the meetings. The grand design initiative opened up a new window of opportunity to garner support for New Markets as a new approach to area development. With the route via the provincial government now apparently a dead end, a new route via the municipalities seemed to open up.

However, the initiative for a grand design soon faltered when the municipalities started to fight about leadership, competencies and jurisdictions. Moreover, the province considered the initiative redundant because similar activities were being pursued by three area committees, which had been implementing provincial rural development policies for years. The municipalities also participated in these area committees, so in the eyes of the province it was not useful to start yet another form of municipal co-operation around the issue of landscape development. The initiative by the municipalities was taken over by the province and the municipalities cancelled their financial investments in Core Area South Limburg. Instead, they used those resources to support the Regional Branding initiative, which was another outcome of the New Markets process.

6.2.6 Beyond the scope of Transforum: regional branding

The initiative by entrepreneurs to develop a regional brand[2] emerged more or less spontaneously during the course of 2006 and developed mostly outside the control of the New Markets organisation. The ideas for a regional brand were developed by entrepreneurs from the New Markets process, who formed the Black Riders Consultation (BRC). Interestingly, of the initiatives discussed so far, this one has been the most successful in terms of support from both public and private actors.

The initiative developed from a discussion about collective marketing needs among entrepreneurs who participated in the workshops and round table meetings of New Markets in Heuvelland. The shift in emphasis in the New Markets process to preserving landscape qualities and building network structures seemed to draw attention away from the marketing problems that the entrepreneurs had identified. The marketing issue was the only issue to which all entrepreneurs could relate, and it was even suggested that a sixth 'new market' be built around this topic and this coalition of entrepreneurs.

According to the BRC, the regional economy was not dynamic enough to be competitive in the European context. Parallel to the provincial government's growing awareness about demographic trends, the BRC believed that the shrinking population and emigration of young people to other parts of the Netherlands further diminished the opportunities for new and existing business to develop. The BRC argued that policies should be aimed at keeping young people in the region by focusing on the attractive living and working environment of South Limburg. To do that, the image of the region would have to be upgraded. The excellent 'work-life balance' of the region had to be promoted.

During 2006, these ideas slowly spread to government officials, who were also increasingly questioning the image of the region. This happened first in a more local context. The city of Maastricht was developing a city marketing strategy to promote the city to tourists, residents and companies. In the summer of 2006 the BRC and experts in the New Markets project came up with the idea of linking Maastricht's city marketing to the Regional Branding idea developed in the New Markets process. Maastricht could profit from the promotion of the wider environment of the city, and the BRC was interested in using Maastricht in the brand because in their view Maastricht was the only internationally well known feature in the region. The BRC coalition was therefore broadened to include representatives from the municipality of Maastricht, but despite initial enthusiasm, the branding and city marketing initiatives stagnated. This triggered the provincial government to intervene and to take over the initiative.

---

[2] Regional branding is the deliberate planning of the image of a region. Marketing focuses on the demands and needs of consumers, but branding is more of a self-selected vision, mission and identity (see Riezebos, 2006).

6. New Markets in Heuvelland

This development should be related to the background of political discussions within the provincial administration. As discussed earlier, in that period the province shifted its attention to attracting young people to support the knowledge economy. Moreover, the National Spatial Strategy forced the province to emphasise the importance of the network of cities in the region instead of individual marketing and promotion activities by each city. Therefore, the branding of the region, specifically related to the South Limburg Urban Network, became a major provincial concern in 2007, as reflected in the Provincial Executive's policy programme of 2007 and the Acceleration Agenda of 2008:

> *We want to create an all-round attractive climate for companies, employees and tourists, which is why we will strengthen our regional branding and use our characteristic Limburgian cultural institutions and media, transboundary cultural economy, and linking major events to economic priorities.*
> (Provincie Limburg, 2007: 20)

> *Clear and consistent promotion of this region is of great importance in attracting international knowledge workers.*
> (Raad van Advies Versnellingsagenda, 2008: 44)

In 2007, the province commissioned LIOF to develop the Regional Branding idea into a business plan. LIOF hired experts and by the end of 2007 a business plan was presented which included a proposal for a public-private organisation that would steer the regional branding and marketing activities of the participating stakeholders. This resulted in the following Regional Brand for Heuvelland:

> *South Limburg is a European region in which people from all over the world come together to make high-quality products. It is a region with a clever balance between working and living. Innovation and high quality go together with joie de vivre. Maastricht is the pearl of the region and resonates with the culture and tradition of the region.*
> (Berenschot, 2007: 3)

The major goals of the regional brand with regard to economic development are: development of the knowledge economy (new industries in the field of the life sciences, medical technology and energy systems), gastronomy, and creating a good investment climate by focusing on the good 'work-life balance' provided by the beautiful landscape of Heuvelland. The province assigned three 'pathfinders' to investigate whether the proposed public and private parties were really interested in participation in a foundation for regional branding and whether they were willing to translate this interest into financial investment. These three pathfinders were influential administrators in the public and private sectors. In 2008, the efforts resulted in the establishment of the Regional Branding Foundation, in which all the municipalities in South Limburg and large enterprises participate.

The prioritisation of regional branding on the policy agenda marks the shift in the provincial government's attention from both the knowledge and 'silver' economies to exclusively the knowledge economy and the related issue of retaining young people in the region. The promoters of the idea of Regional Branding are adamant that the regional brand should not be associated with elderly people.

> *We are not focusing on the 'silver economy' because it is not a new concept in the country. What is the added value compared with the Achterhoek [a region in the east of the Netherlands – ed.]? With this topic we cannot distinguish ourselves from other regions....Providing for older people is a core task of the government, but we should not expect the market to take the lead. The market is concerned with preventing the migration of young people to other regions. Young people are welcome to study abroad, but after that we want them to return to South Limburg. We talk about providing a work-life balance.*
> (A member of the BRC)

This is also stated in the business plan:

> *We are not targeting older people. They are already well provided for, and prefer not to live in a region with an ageing population. The brand should therefore not put too much emphasis on wellness and care, but focus primarily on enjoyment for the active population and for visitors.*
> (Berenschot, 2007: 3)

At the time of writing, the foundation has presented an Implementation Plan for the period 2009-2012, in which Regional Branding is described as an image campaign to improve the reputation of the region and prevent further population decline:

> *The main goal of Regional Branding is to broaden the reputation of South Limburg from a region noted for gastronomy and the good life to the ideal 'work-life balance'. It will promote South Limburg not only as an ideal place for a holiday, but also for permanent residence [by companies and employees].*
> (Stichting Regiobranding Zuid-Limburg, 2009: 9)

6.2.7 Summary

The above shows how during the New Markets process issues developed from 'improvement of the tourism structure' to 'securing and strengthening the environmental qualities of Heuvelland through a vital coalition of economic drivers' to a prioritisation of the Healthy Living in South Limburg initiative, and later to a focus on attracting young people by improving the image of the region: Regional Branding. We can conclude that the initial goal of developing an alternative view on sustainable regional development by focusing on innovative economic activities was ultimately narrowed down to an image campaign. Nevertheless, this idea generated wide support among businesses and government authorities in the region. Regional branding even seems to operate as an attractor for other initiatives in the New Markets project as well as other initiatives in the region. Many respondents view the regional brand as an empty container which needs to be filled with specific projects. Initiatives are attracted by the large financial capital of the Regional Branding Foundation and ideas can be easily incorporated into the 'empty' brand 'Everything Points to South Limburg/Maastricht Region'.

Figure 6.3 illustrates how the New Markets initiatives and coalitions developed and how they are connected to the regional policy agendas. The diagram shows that coalitions within the New Markets process frequently changed and fragmented into several coalitions. Moreover, the agenda of the New Markets process became increasingly narrow, starting from the ambition of contributing to sustainable landscape development and resulting in the establishment of a regional image campaign to attract new businesses and improve employment.

The attempts by the New Markets in Heuvelland project team to give a new impetus to the five new product-market combinations and to combine this with sustainable landscape management turned out to be fruitless in terms of agenda setting for sustainable area development. No lines of communication were established with the provincial government's spatial planning and rural developments departments, and therefore there was no discussion about the policy agendas of those departments. The effect on the economic agenda was limited to prioritising first the Healthy Living in South Limburg initiative and later regional branding. However, as these initiatives fitted in well with the Provincial Executive's existing policy programme and the Acceleration Agenda, or with ideas that were being developed by the provincial government at the same time as the emergence of the initiatives, they did not question or challenge existing policies, but contributed to current debates and 'hot issues' within provincial government. At most, we could say that the initiators of the Regional Branding initiative apparently knew how to adapt to the social trends that would dominate the debate about the new policy programme of the Provincial Executive (demography: the shrinking population and migration of young people to other regions).

That said, the New Markets project did generate a 'buzz' in the region with regard to new product-market combinations: healthcare and tourism ('care hotels' for convalescence following surgery), agriculture and tourism (regional products), housing and preservation of cultural heritage (cloisters and farmsteads), and so on. These themes are still inspiring entrepreneurs in the region to develop businesses. With regard to setting the policy agenda, the result may have been limited to prioritising regional branding, but progress is certainly being made towards concrete economic developments in the region.

In this section we focused our analysis on issue development and how framing of issues influenced the prioritisation of issues on the policy agenda. In the theoretical introduction we stated that agenda setting is not only influenced by issue framing, but also by how support is mobilised through coalition building. The descriptions of the process in the previous section provide keys for the analysis of coalition-building aspects. First, it became clear that issue framing with regard to the contribution of New Markets to sustainable regional development was hampered by the project team's focus on the organisational aspects of the process instead of substantial goals. We reflect on that in Section 6.3. Second, the story of the process shows that the provincial government intervened in many of the events described above, most evidently in the case of the Healthy Living initiative. We will discuss the role of the provincial government in the New Markets process in Section 6.4. The third and final point, discussed in Section 6.5, is how and why the Healthy Living and Regional Branding initiatives, which emerged mostly outside the formal context of New Markets, were successful in setting the agenda.

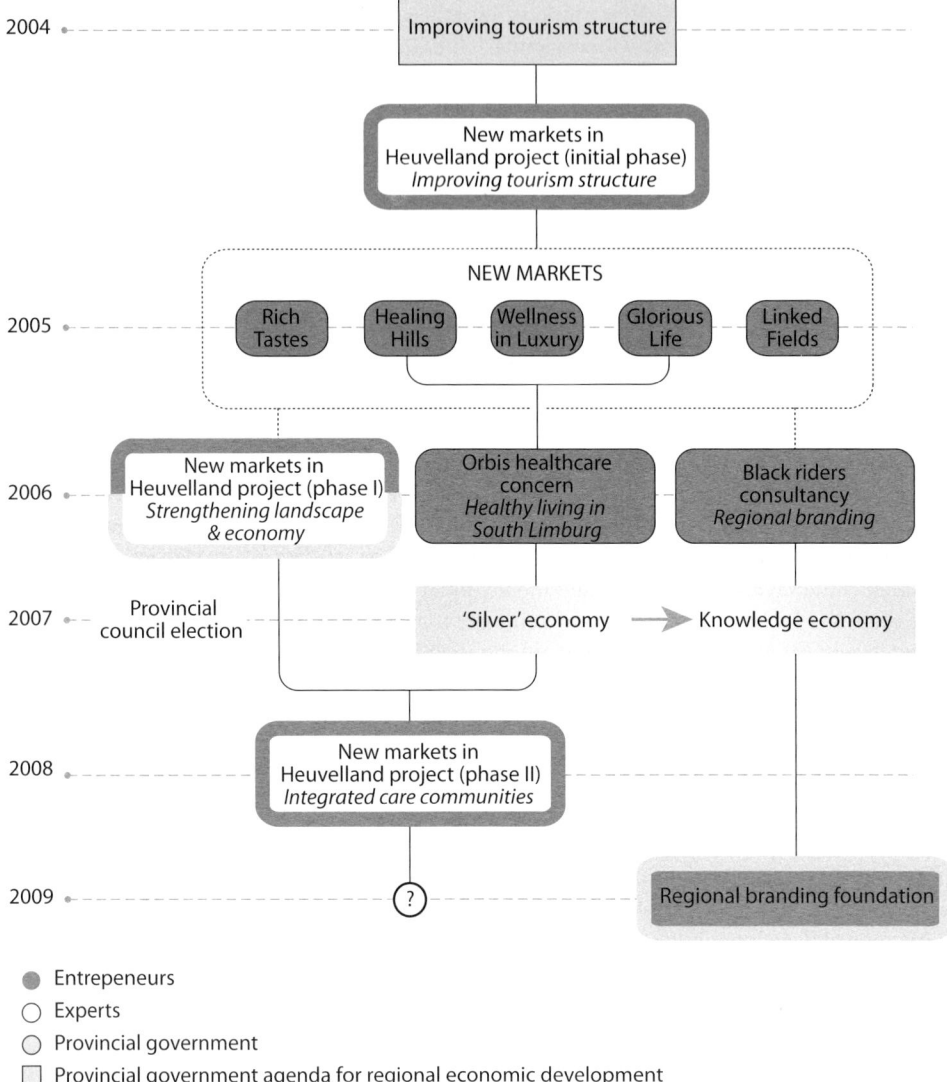

Figure 6.3. Development of the New Markets in Heuvelland project, stakeholders and relation with shifting issues on the provincial government agenda for regional economic development.

6.3 Constructing coalitions without a sense of direction

In Section 6.2 we described how issue framing about the contribution the New Markets initiative could make to sustainable regional development failed because of the project team's focus on the organisational aspects of the process. Transforum and Alterra were included in the process at the beginning of 2006 to give a new impetus to the projects of the entrepreneurs.

Transforum and Alterra set out to make sure that these projects would somehow contribute to the management of valuable landscapes in Heuvelland. However, throughout the process, this adjustment of the initial goal of New Markets was never explicitly discussed, or even put on the agendas of the New Markets meetings.

In this section we argue that the difficulty of defining a clear issue that could contribute to sustainable area development was rooted in the stakeholders' focus on the organisational aspects of the process instead of the substantial goals of the process. We concentrate on four arguments: (1) efforts were aimed mainly at keeping the entrepreneurs in the coalition; (2) experts in the project team argued about who had the power to determine the process architecture; (3) the provincial government pursued its own course; and (4) the project team had little understanding of the informal contexts and power relationships in the decision-making process.

6.3.1 Keep the entrepreneurs committed!

Right at the beginning of the process, the members of the project team were afraid of losing the commitment of the entrepreneurs. They decided that their first priority was to invest energy in getting the entrepreneurs committed to the process. The commitment of the entrepreneurs was considered crucial because the innovative character of the project was attributed mainly to the role of entrepreneurs as initiators. The project was supposed to show that coupled private initiatives are more productive for sustainable area development than a slow multistakeholder negotiation co-ordinated by the provincial government. The enthusiasm and capacity to act of cash-rich entrepreneurs was thought to provide a better solution to sustainable area development. The project managers could therefore not afford to lose the involvement of the entrepreneurs in the coalition.

Consequently, the strategy for the meetings was to find topics that would interest the entrepreneurs: identifying obstacles to the progress of the individual projects, constructing a 'corporate identity' for the region as a business-oriented approach to the qualities of the region, and building a business community and a public-private network organisation to offer a framework in which co-operation between entrepreneurs and other stakeholders could be facilitated. To achieve this last objective, several organisation models were presented and discussed during and between the meetings, but the common goals of such a network organisation and business community were not discussed. Alterra's implicit aim was to construct a network organisation around the issue of landscape management, but this was never put on the meeting agendas. Instead, the emphasis was on the interests of the entrepreneurs, as illustrated by the following quotation, which comes from the minutes of one of the meetings and concerns the mission statement for the New Market project:

> *Entrepreneurs come first! The economic interest has priority and other interests should tie in with this.*

Despite the efforts of the project team, the entrepreneurs in the process did not become more engaged. One of the meetings was even cancelled because the majority of the entrepreneurs pulled out. The project managers noted that the entrepreneurs were frustrated about all the talking without action.

> *I cannot get a clear picture of the benefits to the entrepreneurs of the higher goal of preserving landscape qualities. The project team member from Alterra has developed models, but these confuse the entrepreneurs.*
> (The entrepreneurs' representative in the project team)

Paradoxically, the effect of this desire to keep the entrepreneurs committed and the resulting discussions on how to build a coalition was the withdrawal of the entrepreneurs from the coalition-building process. Interestingly, the entrepreneurs did attempt to co-operate with each other and with the provincial government in informal settings outside the official New Markets process. Some of these activities did lead to successful policy agenda-setting, such as the Regional Branding initiative.

6.3.2 Two captains on the ship

The second reason for prioritising organisational aspects on the New Markets agenda can be found in the conflicts within the project team about the process architecture. The New Markets project management included experts who represented two different schools of thought about the organisation of innovation. The initiator of the New Markets process, who operated as an independent chairman in phase I, believed in selective mobilisation of actors, following the principle that only actors that are really committed to the goal of the process and that can invest resources should be involved. New Markets had started out with that idea of only involving wealthy entrepreneurs and keeping the provincial government at a distance. However, Alterra believed in building a network with broad representation from society, including researchers, experts, government officials and officers, interest groups, entrepreneurs and citizens, arguing that this would generate broad support for the proposed ideas and facilitate creative solutions that combined all the different kinds of knowledge of the participants. Their goal was sustainable landscape management, but this was never explicitly discussed with the provincial government representative and the independent chairman, whose main concern was how to develop the new product-market combinations and create spin-offs, so for them it was logical to concentrate attention on the entrepreneurs.

The project team members and the chairman invested considerable time and energy discussing and arguing these different approaches to deciding who should participate and who should not. The provincial representative and the chairman feared that Alterra's approach would result in an agenda that would be too broad in scope and complicated for the entrepreneurs, who wanted fast decisions and action. Initially, Alterra was given the task of organising the meetings, but the very first meeting was cancelled by the provincial government representative in the project team because he doubted the effectiveness of the approach.

> *The process management is tricky and not clear at the moment. We want to steer but we also want to allow a lot of room for manoeuvre. As long as the organisational aspects have not been decided, we have to take the lead role. We hired Transfontium, but had to step in because their methods and instruments do not work here. We*

thought Transforum would play a supporting role to solving knowledge questions, but they took a leading role. We are not on the same wavelength. The experts focus too much on the instrumental aspects and take too little account of the culture of governing. We were not happy with that, and that is why we took back control.
(Provincial government representative)

From this it is clear that the project team members not only disagreed about the process architecture, but also about who had the right to make the decision. In the midst of this power struggle, there was no discussion about commonly shared concerns or goals.

6.3.3 The provincial government pursued its own course

We noted above that the provincial government questioned the leadership of Alterra and cancelled one of the proposed meetings. Other developments also show that the province pursued its own course and did not wish to invest much energy in the New Markets process.

First, the project manager from Alterra noted, with hindsight, that the provincial representatives never commented on the proposal for the New Markets project, even though the provincial government was formally one of the applicants of the proposal. The provincial representatives said in the interviews that they 'hired' Transforum for the job. The writing of the proposal was seen as a job for Transforum and Alterra, not as something that would matter to the province. Moreover, this attitude suggests that the provincial government never considered Alterra to be a party with authority to redefine the goal and approach of the New Markets project.

Second, the meeting in November 2006 was supposed to be organised by Alterra, but was dominated by the provincial representatives and discussions were for the most part limited to the Healthy Living in South Limburg initiative. The executive councillor for economic affairs appointed the Orbis director as the new project manager for New Markets without consulting the project team. The Alterra project team member refers to this event as a 'coup'. Alterra was excluded from decisions about the process architecture and desired outcome.

Third, although there were some attempts to involve the provincial government's rural development and spatial planning departments, they never seriously considered New Markets to be a promising initiative for landscape management. They pursued their own agenda and were not inclined to consider initiatives that were not immediately relevant to existing policy goals. The initiator of the New Markets process tried to link the New Markets vision to the spatial planning department's objective of drawing up a Landscape Vision – the timing was right to couple these two processes as both were at an initial stage – but the spatial planning department claimed they first needed to design their own vision for the landscape before they could adopt a position on the New Markets process. The initiator calls this a 'not invented here' attitude. Later on in the New Markets process, when Transforum introduced the broader goal of investing in landscape qualities, the project team wanted to involve representatives from the spatial planning and rural development departments again. However, according to project team members, the officers from these departments and the economic affairs department were unwilling to find ways to co-operate.

The interviews show that the spatial planning and rural development departments hardly knew what New Markets was: 'something with healing hills' or 'something to do with health, care and cure'. Provincial government officers felt that the initiatives were 'too vague', which made it difficult to see the added value for the province. 'All beautiful terms, but I don't feel it, it doesn't click'. The policy frame for the spatial planning and rural development departments dictates that projects and implementation of policy should be initiated by stakeholders in the region. Once initiatives have been firmed up, the province then tests whether they fit in with the goals set down in the policy programmes for spatial planning and rural development. The New Markets initiatives were considered not to be detailed enough to fit in with the specific goals of the policy programmes, which made them unattractive for serious consideration by the spatial planning and rural development departments.

6.3.4 Informal networks prevail

A paradoxical situation emerged regarding the attempts of the project team to construct a stable coalition around New Markets. While discussions in formal meetings focused on how to build such a coalition, stakeholders met in different informal settings and in different groups to discuss substantive matters that really interested them. This resulted in the BRC initiative for Regional Branding. Although the project team never had a complete picture of the negotiations that took place behind the scenes in these smaller informal groups, these informal discussions were decisive for later decisions about the New Markets process, according to the project manager from Alterra:

> *With regard to the process I am not happy with all these individual discussions. There is no context, no sense of cohesion. Here in South-Limburg, one-to-one relations are considered more important than collective discussion. You belong here, or you do not. If you do not belong, you are excluded unless you have a formal position. You will get an assignment if you speak Limburgish. It is very much an in crowd. Everything happens in backrooms, like the Black Rider Consultation.*

In the view of respondents, people in the regions prefer to arrange matters in informal one-to-one discussions. The stakeholders preferred to work in smaller coalitions. In such a situation, it was difficult to co-ordinate a collective negotiation and to construct a storyline or issue that could count on the backing and commitment of so many parties. Conversely, the agendas of the New Markets meetings, which focused mainly on organisational aspects, provided no incentive to discuss substantive issues as a common ground for co-operation.

6.3.5 Conclusion

This section shows that issue framing and coalition building are intertwined agenda-setting mechanisms. Lacking a good storyline about the contribution of the New Markets to sustainable regional development, the project team was not able to make other actors enthusiastic. This made it difficult to construct a coalition of actors with the right resources to help develop and lobby for a new perspective for sustainable regional development. The findings suggested four reasons for the failure to construct an issue, all related to coalition-building aspects of agenda setting. The first two concern the activities of the project team: the focus on building a network structure and assembling a business community, and the conflicts between the experts about the process architecture. The second two reveal how the provincial government and the informal contexts surrounding the New Markets initiative hampered the construction of a clear issue. In this section we have seen how the provincial government and informal contexts played an important role, but have only discussed this in relation to the failure to construct an issue in the New Markets process. In the following sections, we further reflect on the role of the provincial government and informal contexts in coalition building and relation framing.

6.4 The 'constructed power' of the provincial government

The previous sections revealed the important part played by the provincial government during the course of the New Markets process in determining the issues that were or were not addressed and the outcomes. In relation to the construction of coalitions, we now turn our attention to the perceptions about the role played by the provincial government. We discuss why the stakeholders attached so much importance to the participation of the provincial government in the New Markets process and draw some conclusions on how the 'constructed power' of the provincial government – the power or authority attributed to it by people in the region – influenced the framing of issues by other actors so that they tied in with existing economic policies. The power of the province is thus continuously reaffirmed and strengthened.

6.4.1 Constructing the role of the provincial government in the New Markets process

Non-governmental stakeholders in the New Markets process held ambiguous views about the role of the provincial government (for similar experiences see Aarts *et al.*, 2007). On the one hand, stakeholders claimed that they wanted to remain independent of the provincial government, but on the other hand their statements show that they wanted the province to take a much more active stance.

The most frequently raised complaint is that the provincial government was too passive and did not take the initiative. In Section 6.3, we saw that the province assumed that other regional stakeholders, such as local authorities, farmers, water boards, etc., would take up and implement provincial policies. Project managers and entrepreneurs saw the provincial government as an institution that was only interested in controlling processes and not stimulating new developments, assessing initiatives against their applicability to existing policy programmes, but not helping to develop ideas or come up with new ideas itself. In some cases the province adopted initiatives started by others and incorporated them into current institutions, thereby destroying the innovative potential of those initiatives. Another obstacle to co-operation in the eyes of the respondents was the compartmentalisation within the organisational structure of the authority. It was described as a 'splitting zone' because of the lack of willingness of the economic affairs department and the rural development and spatial planning departments to share ideas and co-operate in joint projects. This not restricted to communication between provincial government officers, but also between the executive councillors, who were said to act too much on their own and not consult other stakeholders.

The complaints about the provincial government suggest that it was not considered to be trustworthy. Nevertheless, the involvement of the province in the New Markets process was thought to be necessary to get results. Stakeholders argued that the province should 'just facilitate' the initiatives of entrepreneurs and 'nothing more'. The members of the project team were afraid that too much government involvement would suffocate the process by strangling creativity and productivity. They suspected that if the initiative became dominated by provincial government, it would become institutionalised, thereby diminishing its innovative potential. However, 'facilitation and nothing more' turned out to mean more than the words suggest. Stakeholders expected the province to be more helpful with setting up facilities for the new businesses that were proposed, especially by providing information about the relevant regulations and financing possibilities. Moreover, some think that the province should be responsible for the co-ordination between entrepreneurs, both in terms of organisation and by 'providing a common theme' for ideas and initiatives.

> We would like to do something together, but then according to one plan, so no fragmentation please. Province, coordinate that! Make sure it is arranged. But the provincial government lacks the necessary continuity or capacity and organisation. Entrepreneurs are usually not interested in institutions, but in people who are entrepreneurial like themselves, so that they can get going.
> (The entrepreneurs' representative in the project team)

If the provincial government operated in this way, it would steer initiatives much more than the term 'facilitation' suggests. By focusing on the 'leading' role of the entrepreneurs the organisers of the New Markets process tried hard not to make the province the key actor. However, when we look at their expectations and attempts to involve provincial government officers from different departments and the relevant executive councillors, they were in fact affording the provincial government a central role. In any case, confusion and disappointment about the role of the province led to alternately investing energy in involving provincial representatives in the process and then pushing them out again.

6.4.2 The power of the provincial government: reproduced in conversations

Despite what the other parties said they wanted, the provincial government was therefore given a central role in the New Markets process, although often implicitly and with some reluctance. Apparently, it was not considered possible or desirable to try to sideline the provincial government. The interviews suggest that in general stakeholders perceive the provincial government as a powerful and decisive actor in the region, which explains the importance placed on its involvement in projects. We support this contention with three arguments: (1) the emphasis on the power of individual executive councillors; (2) the constructed power of the province in general; and (3) the ability of the province to use a culturally defined focus on influential people, including informal ways of communication in changing networks.

First, respondents claim that people in the region rely on the power of individuals to influence decision-making and negotiation processes. The executive councillors of the economic affairs department and the rural development and spatial planning departments of the province are considered to be such influential persons. They are frequently called 'kings' of the regions and also view themselves as people who can make a difference, as this statement by a former member of the Provincial Executive shows.

> *In Limburg, regional officials are influential. That cannot be said everywhere else. We do not have a big city in Limburg, which means the province has more influence. Every day there is something about the province in the newspaper or on the television. People know who their councillors are. They elect people who appeal to them. Although this is a national trend, it is more evident in Limburg than in other provinces. People want personal contact, empathy, an experience. People make the difference.*
> *We Limburgians are sensitive to authority and status; what those people say, goes.*
> (One of the participants in the New Markets process)

The influence of individuals is viewed as a cultural characteristic of the region, as are the informal one-to-one negotiations. Furthermore, respondents argued that people born and raised in the regions have more authority.

Second, respondents said that people in the region in general have a noncommittal attitude towards the provincial government. This attitude is said to originate from a collective lack of self esteem and the 'guts and spirit' needed to take the initiative. This lack of self esteem can be explained by historical reasons, such as the isolated position of the region in relation to the rest of the Netherlands and the traumatic period of unemployment after the closure of the coal mines. This is illustrated by the following statement by a past executive councillor of the provincial government.

> *Certain qualities are lacking in the region: self-awareness, pushing back frontiers, identifying opportunities, not complaining, taking matters into one's own hands. The region looks to The Hague too much, and does not want to get things done itself. We are the second youngest province in the Netherlands. We never really belonged to the Netherlands. We were incorporated into the Netherlands in 1839, but we were also a member of the German Confederation for a long time. Actually, during our history we have belonged to everybody at one time or another and have been used as a pawn in international deals. We have therefore never developed a sense of belonging and have never had the guts and spirit to make our own choices for the future of the region.*
> (A former member of the Provincial Executive)

This reference to a collective lack of willingness to act among regional stakeholders puts the authoritative role of the province into perspective. Expectations about who should take the initiative take two forms. On the one hand, people complain about the noncommittal attitude in the province, and on the other hand non-governmental stakeholders in the region are also characterised as being reticent and lacking the 'guts' to take matters into their own hands.

A third reference to the power of the provincial government concerns its awareness of the important role of informal decision contexts and the ability of the provincial government to operate in such contexts. Respondents suggest that networks cannot be identified as fixed groups of people. Coalitions and networks change continuously, which means that the networking strategies of the province have to adapt to the shifts in those networks.

> *It is not possible to organise cooperation in a rational way. I think this is especially the case in Limburg. Here you depend a lot on political and administrative relations: 'who knows who'. The political ambitions of some determine whether others will be accepted or removed from a particular position. That always happens at eight o'clock in the evening. All the telephones in the province are ringing then. That is when business is done in Limburg. It's all rather intangible.*
> (A provincial government officer)

Co-operation in networks is clearly perceived to be intangible, which makes it impossible to define strategies beforehand. Nevertheless, the interviews show that the provincial government does consciously try to influence co-operation in networks. Ultimately, the province will decide on what is synergetic co-operation, what the conditions are for useful co-operation and which types of networks should be given more power. Moreover, the province is considered to be the only actor that has an overview of what happens in the region in terms of activities and networks.

> *Sometimes, as in the 'core area' of Heuvelland, a separate steering group emerges [the initiative by the municipal councils to form the Core Area South Limburg network and develop a 'grand design' – ed.]. This group turned out to be doing the same things as the area committees, which is why the 'grand design' was incorporated into those area committees, resulting in only one structure....That is how we try to organise useful cooperation.*
> (A provincial government officer)

We argue that the emphasis on informal contexts can be inferred from the previously described preference for relying on influential individuals and not on rules and programmes. One-to-one communication between individuals is considered more decisive than negotiation in formal settings. The province knows how to use this situation for its own ends by appointing individuals to arrange for co-operation and to connect networks and ideas, as illustrated by the following quotation.

> *The Provincial Council approached me because of my efforts in organising cooperation between seven municipal councils and setting up a regional authority for Parkstad. I saw that the problems of the region could not be solved if the municipalities each tried to pursue their own interests. That is probably why the province approached me. They saw me as a fighter who perseveres to the bitter end, not a quitter. Another reason is that I was born and raised here and I have put all my energy into this area. I have faith and I do not give up easily.*
> (A former local government officer)

6.4.3 Conclusion

We conclude that the focus on the involvement of the provincial government in coalitions around the New Markets process can be explained by a general tendency among regional actors to ascribe considerable authority and power to the provincial government. This seems to be a cultural characteristic of the region. The power of the provincial government can explain why the province was such a decisive factor in determining the issues in the New Markets process (see Sections 6.2 and 6.3). In Section 6.3 we saw that in relation to the New Markets process the province pursued its own course, which is understandable in the light of the findings in this section: the provincial government is afforded considerable authority by stakeholders in the region as well as its own provincial representatives. In terms of agenda setting, this leads to the conclusion that the province, acting out its dominant role in the region, will only prioritise initiatives that fit into its own ideas, visions and plans. In this section we saw that this dominant role of the province is accepted and reconfirmed by actors in the region. In the New Markets process this is illustrated by a noncommittal and uncertain attitude towards the province, and also by how issues were framed in terms of existing provincial government policy frames in order to gain support.

Next, we reflect on why the self-organised initiatives of Healthy Living in South Limburg and Regional Branding were successful in setting agendas, compared with the failed attempt of the broader New Markets organisation to set an agenda for sustainable area development.

6.5 Self-organisation and synchronicity

During the New Markets process two initiatives, which developed more or less autonomously and outside the grasp of the New Markets organisation, turned out to be the most successful in setting agendas: Healthy Living in South Limburg and Regional Branding. Both initiatives were taken by entrepreneurs to create favourable business conditions for themselves. In doing so, they served the objectives of the provincial government. We discuss two factors which explain their success: the ability to connect formal and informal networks, and the ability to exploit synchronicity.

6.5.1 Connecting formal and informal contexts

The research findings suggest that the priority given to the Healthy Living in South Limburg initiative on the policy agenda had little to do with the efforts of the project managers of New Markets. To the Alterra project manager, it came as a surprise that the November meeting was primarily devoted to this initiative and that the healthcare concern director was appointed as project manager for the second phase of New Markets by the executive councillor for economic affairs. This 'coup' by the executive councillor was seen as something that occurred outside the reality of the New Markets process.

> *During the process the representatives of the provincial government's economic affairs department wrote reports, but I haven't seen them and nothing was done with them. The executive councillor for economic affairs did not know about the*

> involvement of Transforum in the New Markets process. His interest was that the project would be in good shape at the end of his term. That is why he approached the Orbis director.
>
> (Alterra project manager)

The project managers also talk about how the executive councillor for economic affairs was not interested in attending the New Markets meetings. This suggests that he was more interested in concrete business outcomes than in the attempts of the New Markets process to create a broader network organisation with representation from several provincial departments, interest groups and entrepreneurs. Moreover, the executive councillor was apparently convinced that the healthcare concern director had the leadership qualities required for the New Markets process. Orbis Healthcare Concern had contributed to the Health, Care & Cure theme of the Acceleration Agenda and had therefore participated in the coalition of entrepreneurs and government authorities assembled to establish the Acceleration Agenda. The activities of Orbis and the Orbis director already had the attention of the executive councillor before the ideas for Healthy Living were proposed.

The development of the initiative for Regional Branding also illustrates the importance of connecting informal and formal networks. The Black Riders Consultation arose from informal contacts between entrepreneurs who met in the New Markets process and were interested in collective marketing needs. As described in Section 6.3, the New Markets process did not result in a clear issue formulation and was mainly limited to a discussion about building a network organisation. Although the organisers intended to discuss matters that would attract the entrepreneurs, the result of their efforts was in fact a drop in commitment from those entrepreneurs, who then pulled out of the formal process and found a way to organise themselves in an informal setting. One of the entrepreneurs in the BRC explains how he brought other entrepreneurs from the New Markets process together in informal meetings:

> *The BRC are people from the round table meetings in the New Markets process. I regularly met them there and it turned out that we shared many of the same views about the lack of action and cooperation. The difference was that they were born and raised here and know the ins and outs of this area. So I brought these people together to find out if there could be an initiative for business interests. We considered who should be involved and who not.... We were invited by the former director of the Gulpener brewery to hold the first meeting in his pub 'The Black Rider'. At the end we needed a name: the Black Riders Consultation.*

The group consisted of successful entrepreneurs, including some who were active in governmental contexts. Together they had a network of about fifty successful entrepreneurs. Their connections with the network of government officials and how their ideas connected with the current economic policy agenda (see Section 6.2) can explain why in 2007 the province decided to help develop the Regional Branding initiative. Three 'pathfinders' were assigned to co-ordinate the development of a business plan and find support among entrepreneurs and municipal councils. One of them was a member of the Black Riders Consultation who had earlier proved his ability to combine business interests with meeting the objectives of the provincial government. Another, was a former successful businessman and politician, and the third was a government official with firm roots in the political and policy context. Their efforts led to the establishment of the Foundation for Regional Branding. This shows that right from the start the informal group of entrepreneurs had firm roots in the informal network of public and private executives and were able to build on those ties in the pursuit of the Regional Branding initiative.

6.5.2 Synchronicity influences agenda setting

An important agenda-setting condition in Kingdon's work (Kingdon, 2003) is formed by political or social 'trigger events', which provide an opportunity to focus the attention of policy makers on a specific issue (window of opportunity). This suggests that agenda setting may depend on coincidence. However, we argue that issue proponents or policy entrepreneurs apparently interpret concurring events as meaningful and useful for the issue they are trying to push onto the policy agenda. Therefore, we would rather speak of a concurrence of circumstances or *synchronicity* (a term coined by Carl Jung for the apparently meaningful coincidence in time of two or more similar or identical events that are causally unrelated). Below we discuss the phenomenon of synchronicity in the agenda-setting process for New Markets in Heuvelland.

Initially, the Healthy Living in South Limburg initiative seemed to fit in very well with the goals of the Provincial Executive in office at that time, as described earlier, and it was successfully put on the economic policy agenda. However, only three months later it was dropped from the agenda and the Regional Branding initiative was given priority, as described in Section 6.2. We can explain this by two concurrent events: (1) the emerging awareness among stakeholders in the region of the consequences of demographic trends for the economy, initially focusing on the ageing population but over the years shifting to the problem of the loss of young people from the regional labour market, and (2) the Provincial Council elections in 2007, which provided the opportunity for policy changes. The Healthy Living and Regional Branding initiatives were both well connected through the informal and formal networks of business and government executives. This would suggest that in this case successful issue framing and the ability to adapt to synchronicity was decisive for agenda setting. Apparently, it was crucial to frame ideas into issues that would be important to the provincial government for the longer term. To do that, initiators had to be able to respond to concurrent events, the slow demographic trend, the political debates and the fact that the provincial elections were near.

6.5.3 Self-organisation reproduces existing frames of the provincial government

The Healthy Living in South Limburg and Regional Branding initiatives are both self-organised groups of entrepreneurs who operate in informal and formal policy networks which are highly connected. In fact, we can state that we are dealing with a network of influential executives from the public and private sectors in which informal and formal interaction are closely connected. We showed how interaction, whether informal or formal, resulted in the re-establishment of provincial policy goals for both economic development and landscape development. The province was framed as the body that has most authority and capacity to act in regional development, and therefore the policy the Provincial Executive's policy programme was accepted as a guideline for future action. This is the case for both nature and landscape policy and economic policy. The Healthy Living and Regional Branding initiatives viewed the spatial planning and rural development policies as given and did not question them or try to discuss them with representatives from the provincial government's spatial planning and rural development departments. With regard to the economic policy, we saw that the focus shifted from the 'silver economy' to the knowledge economy. This change was brought about by the increasing socioeconomic problems posed by demographic trends and the strategies that were gradually developed by the network of provincial government officials and business executives in the region. The Regional Branding initiative was not revolutionary in that context, but strengthened the ideas that had already been developed. In any case, both the 'silver economy' and the knowledge economy can be traced back to a desire of the province to create an economically competitive region in the international context and move on from the economic traumas of the past.

We can conclude that the initial goal of the New Markets process – to develop and promote a new approach to area development – was not achieved. Instead, the actions of self-organised groups led to a re-establishment of economic policy goals. The ambition of the New Markets process was narrowed down from sustainable area development to an image campaign for the region.

6.6 Conclusion and discussion

In this chapter we have shown the capricious process by which the New Markets initiatives have developed agenda-setting capacity. We focused on issue framing and the formation of coalitions around issues. We discussed how these activities affected regional policy agendas for sustainable area development. So what can we learn from this about creating agenda-setting capacity as a condition for vital coalitions?

First, we can conclude that a clear direction, in the form of an appealing story, is of the utmost importance. In the New Markets process it was not possible to formulate an appealing and directional ambition for area development. There was too much focus on building coalitions and managing relations, and too little on identifying issues. On the other hand, identifying issues in a collective setting was difficult because stakeholders preferred to realise their own ambitions in smaller and informal settings. As a result, the New Markets initiative, as an alternative approach to area development, was not able to develop an appealing storyline that would generate attention and broad support for the proposed ideas. The lack of direction was compounded by a coming and going of stakeholders. The coalitions seemed to be unmanageable and directionless.

Second, vital coalitions are able to connect to relevant governmental actors, in this case in provincial government. Stakeholders perceive the provincial government to be a crucial and powerful actor in regional development. Therefore, agenda-setting strategies were mainly aimed at the provincial policy agendas and at getting the attention of provincial representatives.

Third, vital coalitions should look for initiatives with high self-organising potential and try to connect with those initiatives. In the New Markets process, two initiatives developed outside the formal context of the process, but these turned out to be the most successful. One success factor is their connection with the broader informal and formal network of public and private executives in the region. A more decisive factor was the ability to frame issues in terms of goals that are recognisable for the province in the longer term, and related to that the ability to exploit synchronicity (political, social, etc.).

6.6.1 Consequences for vital coalitions and regimes

The New Markets process was used as a case study to get more insight into the question of how vital coalitions can be established between public and private stakeholders for sustainable area development. We focused this chapter on the agenda-setting capacity of the New Markets initiatives and concluded that initiatives lacked the capacity to formulate a clear and appealing ambition. In any case, without a clear issue, it was extremely difficult to generate support from other actors and create stable coalitions.

We may, however, contest the need to create stable coalitions. The respondents consider instability and fluidity of networks to be a fact of life in South Limburg and adapt their strategies accordingly. The provincial representatives attempt to organise co-operation which in their eyes is useful, but they realise that political and managerial relations are intangible so they have to adapt to continuously changing situations. Relations and circumstances change all the time, so coalitions will change as well. If coalitions never changed, there would be no stimuli for fresh ideas or new solutions to changing situations; initiatives could become locked in and slowly disappear. Movement within coalitions is needed to drive the motor of the innovation process.

However, in addition to movement in coalitions, a clear direction is also needed, as this case study has showed. If there is no appealing story to tell, if there is no real issue that is recognisable to potential supporters, then there is no direction for the actions of coalitions. This suggests that the four dimensions of the vital coalition concept – agenda, actors, modes of alignment, resources (see Chapter 3) – are not equally important for the vitality of coalitions. We argue that more emphasis should be put on the 'agenda' dimension of the vital coalition concept. The condition for vital coalitions can be formulated as *constructing a powerful storyline that is easily passed on to others*. In the context of South Limburg, in which stakeholders rely heavily on the authority of the provincial government, such a storyline had to connect with existing policy frames. As a consequence, the innovation potential in South Limburg was not very large. In this context, change depends on the ability of the provincial government to open up to new perspectives.

So what does this mean for the concept of vital coalitions? We suggest viewing the concept more as a process rather than a fixed group of stakeholders. Vital coalitions would then be a process of coalition building in which the creation of a powerful and appealing storyline is more important than holding on to participants in the coalition. The storyline should be strong enough to survive its creators. A vital coalition can thus be reformulated as *a changing coalition or a conglomerate of coalitions that carry a durable and appealing storyline which connects ideas, resources and people*.

References

Aarts, N., C. Van Woerkum and B. Vermunt, 2007. Policy and Planning in the Dutch Countryside: The Role of Regional Innovation Networks. Journal of Environmental Planning and Management 50(6): 727-744.
Berenschot, 2007. Businessplan Regiobranding Zuid-Limburg. Berenschot Utrecht, the Netherlands.
Dearing, J.W. and E.M. Rogers, 1996. Agenda-setting. Sage: Thousand Oaks, CA, USA.
Dewulf, A., B. Gray, L. Putnam, R. Lewicki, N. Aarts, R. Bouwen and C. Van Woerkum, 2009. Disentangling approaches to framing in conflict and negotiation research: A meta-paradigmatic perspective. Human Relations 62(2): 155-193.
Entman, R., 1993. Framing: Towards clarification of a fractured paradigm. Journal of Communication 43(3): 51-58.
Entman, R., 2003. Projections of power: framing news, public opinion, and U.S. foreign policy. University of Chicago Press: Chicago, IL, USA.
Hajer, M.A., 1995. The politics of environmental discourse: ecological modernization and the policy process. Clarendon Press: Oxford, UK.
Horlings, I., and W. Haarmann, 2008. Vitale verbindingen, botsende belangen. Positionpaper over het project Vitale Coalities en Nieuwe Markten Heuvelland. In: Transforum, Innovatief Praktijkproject Nieuwe Markten en Vitale Coalities Heuvelland Zuid-Limburg, Workingpaper no. 8. Transforum: Zoetermeer, the Netherlands.
Jones, B.D., and F.R. Baumgartner, 2005. The politics of attention: how government prioritizes problems. University of Chicago Press: Chicago, IL, USA.
Kingdon, J.W., 2003. Agendas, alternatives, and public policies. Longman: New York, NY, USA.
McCombs, M.E., and D.L. Shaw, 1993. The evolution of agenda-setting research: Twenty-five years in the marketplace of ideas. Journal of Political Communication 43: 58-67.
Mintrom, M., 1997. Policy entrepreneurs and the diffusion of innovation. American Journal of Political Science 41(3): 738-770.
Mommaas, H. and J. Janssen, 2008. Towards a synergy between 'content' and 'process' in Dutch spatial planning: the Heuvelland case. Journal of Housing and the Built Environment 23(1): 21-35.
Provincie Limburg, 2007. Coalitieakkoord 2007-2011. Investeren en verbinden. De mens centraal in een vertrouwde omgeving. Provincie Limburg: Maastricht, the Netherlands.
Raad van Advies Versnellingsagenda, 2008. Versnellingsagenda 2008-2011. Naar een hogere versnelling. Programmabureau Versnellingsagenda: Maastricht, the Netherlands.
Rein, M. and D. Schon, 1996. Frame-Critical Policy Analysis and Frame-Reflective Policy Practice. Knowledge and Policy 9(1): 85-105.
Riezebos, R., 2006. City Branding: zin of onzin? BrandCapital/EURIB: Rotterdam, the Netherlands.
Scheufele, D.A., and D. Tewksbury, 2007. Framing, agenda setting, and priming: The evolution of three media effects models. Journal of Communication 57(1): 9-20.
Stichting Regiobranding Zuid-Limburg, 2009. Samen bouwen aan een sterk merk. Uitvoeringsplan Regiobranding 2009-2012. Stichting Regiobranding Zuid-Limburg: Maastricht, the Netherlands.
Taskforce Versnellingsagenda, 2005. Versnellingsagenda: Limburg op weg naar 2012. Provincie Limburg: Maastricht, the Netherlands.
Van Mansfeld, M., and H. Van der Stoep, 2008. Procesverslag innovatief praktijkproject Nieuwe Markten en Vitale Coalities Heerlijkheid Heuvelland fase 1. In: Transforum, Innovatief Praktijkproject Nieuwe Markten en Vitale Coalities Heuvelland Zuid-Limburg, Working paper no. 8. Transforum: Zoetermeer, the Netherlands, pp. 19-94.
Weaver, D.H., 2007. Thoughts on agenda setting, framing, and priming. Journal of Communication 57(1): 142-147.
ZKA Leisure consultants and planners, Urban Unlimited urban and regional planners, University of Tilburg department of leisure studies, 2005. Heerlijkheid Heuvelland; nieuwe markten en allianties voor toerisme in het Heuvelland. Authorised by LIOF: Maastricht.

Chapter 7
Conditions for vital coalitions in regional development

Julien van Ostaaijen, Ina Horlings and Hetty van der Stoep

7.1 Introduction

The Dutch countryside is in transition. In some regions a variety of closely interrelated environmental, social and regulatory problems are bringing about a process of rural change. These regions are being transformed from predominantly production-oriented zones towards consumption-oriented ones. A declining number of agricultural entrepreneurs see opportunities to scale up and specialise their production, while others sell their farms or switch to multifunctional agriculture. In densely populated parts of the country, agriculture is giving way to new functions and activities, which are creating a new social and physical environment.

Boundaries between cities and the countryside are disappearing. Rural space in the Netherlands can even be considered as having an urban function because the needs of the city largely determine what happens in the countryside. Besides the growing demand for housing land, space is also needed for water retention and flood protection, the development and restoration of natural habitats, and recreational areas and facilities to meet the demands of urban dwellers for space, peace and quiet outside the city. As the traditional urban-rural divide becomes increasingly blurred, these areas are better described as 'metropolitan landscapes', which does more justice to the close relations between cities and their environment (Van der Valk and Van Dijk, 2009). The traditional distinction between urban and rural spatial policies, each with their own specific group of experts, policy makers and stakeholders, does not match the new challenges of the changing metropolitan landscapes.

Many regions are looking for new orientations to solve the current dilemmas of regional planning and transition. In urban planning and policy analysis the concepts of urban regime and vital coalition have been used to study such dilemmas and developments in the local urban context. The concept of vital coalitions has been used to describe and analyse how small-scale initiatives can contribute to urban development. In a similar fashion, regimes are used as conceptual tools to analyse the governance of urban development at the decision-making level.

Our research set out to discover what insights could be obtained by applying these concepts to the regional level. Are there initiatives and co-operative ventures at the regional level that can be described as vital and that aim to contribute to sustainable regional development? If so, what can we learn from them, and what recommendations follow from these conclusions? When translating these concepts to the regional level, we are especially interested in how vital coalitions interact with regional regimes. In other words, whether there are regional structures or forms of governance that stimulate or hamper such coalitions. This led to the following central research questions, which were introduced in Chapter 1.

> *How vital are public-private coalitions that aim to contribute to sustainable innovations in rural-urban regions in the Netherlands?*
>
> *How do current regime conditions hinder or facilitate the mobilisation of vital coalitions?*
>
> *What new conditions are needed to organise vital coalitions?*

In this chapter, we assess the results of our research and present the insights it provides to help answer these questions. First, we reflect on how several coalitions have expressed themselves at the regional level, and in what way they can be called vital. We then focus more explicitly on the conditions for vital coalitions at the regional level before elaborating on the use of the vital coalition and regime concept. Finally, we present some recommendations for the organisation of vital coalitions within regional development.

7.2 Vitality of the studied 'vital coalitions'

In the previous chapters, we defined vital coalitions as a form of active citizenship and self-organisation in which citizens and/or private or public actors take the initiative to act on behalf of a common concern or interest. This book has covered many examples of such forms of private or public-private partnerships. Horlings (Chapter 4) has identified three different types: *business-oriented coalitions* between different agricultural or non-agricultural sectors, *rural urban alignments* between entrepreneurs and consumers/citizens, and *steering coalitions* between entrepreneurs, NGOs and government.

There are several similarities between these coalitions in terms of the four dimensions of regional regimes identified from urban regime theory in Chapter 3: agenda, coalition, modes of alignment and resources. With regard to agendas, most of them have or have attempted to introduce *new strategies* which can contribute to sustainable development, such as regional branding, regional farm management, new organisational arrangements, clustering of businesses, new regional designs and alliances with non-agricultural sectors. The actors in *coalitions* generally include not only citizens and entrepreneurs, but also governmental actors and knowledge institutions. *Modes of alignment* are both formal and informal in nature and often depend on the culture of the region and the specific coalition that is being studied. Informal alignment is important in offering the possibility of direct contact with government executives and of negotiations behind the scenes outside the formal setting. Finally, the cases showed how a wide range of *resources* are brought into coalitions to accomplish agendas, including time, knowledge, energy, money and/or materials.

These similarities describe a set of characteristics of the studied coalitions in the context of sustainable regional development. However, this does not answer the questions about the vitality of these coalitions and the conditions required for their vitality.

7. Conditions for vital coalitions in regional development

Vital coalitions are characterised as energetic and productive. Throughout this research we have deepened this characteristic by looking at vitality in two different ways. On the one hand vitality can be evaluated by analysing whether coalitions have achieved results and have been able to 'make a difference' in regional development; in other words, whether they have been able to build capacity to act. On the other hand, vitality can be evaluated as the more intangible outcomes of the efforts made by coalitions. Chapter 5 operationalised vitality by identifying factors such as motivation, passion, inspiration and energy: qualities that enable initiators and supporters to maintain the momentum required to achieve their goals. In that sense, vitality is something that emerges in the process and is a by-product of the actions of the participating actors and the interactions between them.

With regard to vitality as a condition for achieving results, the findings show diverse outcomes of the researched coalitions. Most of the projects we investigated have influenced local or regional agendas and have been able to create some room for manoeuvre and realise their goals, at least in part. Their contribution to sustainability on a regional scale has remained somewhat limited so far, especially in terms of social aspects (the People dimension of sustainability). The Sjalon, Het Groene Woud and New Markets projects also show that it is a difficult and time-consuming process to scale up innovative projects to the regional level, to align different actors, co-ordinate story lines and to seduce entrepreneurs to invest in public goals such as landscape quality based on new market opportunities.

With respect to evaluating the results of coalitions, we were also interested in whether the coalitions changed anything in the 'rules of the game' and in the dominant policy frames of current regimes. Some of the coalitions have influenced policy agendas, such as the branding initiative in Het Groene Woud, the farmers' plan in the Overdiepse Polder, and the new city-countryside relations in Waterland. However, changing the institutional rules seems to have been too ambitious a goal in most cases. The most striking example is the Northern Frisian Woods case, in which a coalition of farmers have been trying to obtain a formal experimental status for their activities since the beginning of the 1990s, but have not yet succeeded in lifting all the constraints imposed by the environmental regulations.

The diffusion of innovative ideas proposed by the studied coalitions at the regime level seems to be hampered by the following regime aspects:
- *Organisational policy constraints and implementation problems*: the fragmented provincial governmental organisation, the rural-urban divide in governance, and the tension between the situational logic of entrepreneurs and the institutional logic of governments (a specific expression of the last problem is the impatience of private actors when faced with the different time horizon of public actors).
- *Iinstitutional obstacles*: the mismatch between general laws and regulations at the national and European levels and the specific situation or regional characteristics.
- *Clashes between different regional agendas*: the agendas of vital coalitions do not always connect well with the agenda of the 'regime' or regional actors in general.

Faced with these obstacles, innovative entrepreneurs have to invest time in gaining political support from government authorities on different levels, which involves dealing with existing, and sometimes additional, rules and regulations. We elaborate more on the limits and constraints imposed on vital coalitions in Section 7.5.

Besides regime dynamics (see also Section 7.5 below), other factors can also reduce vitality, such as difficulties in aligning entrepreneurs behind a common cause. This is clearly shown in the Sjalon case, in which entrepreneurs were reluctant to give up ownership of their land and machinery. Obstacles to forming a business community were also encountered in the New Mixed Business and New Markets in Heuvelland cases.

With regard to vitality in terms of the intangible process outcomes of energy, motivation and inspiration, we can conclude that most energy and inspiration can be found in small bottom-up coalitions that start as private initiatives and are based on a shared private and common concern held by people who know the region well. The role of entrepreneurship and leadership, often embodied in one or a few individuals, proved to be important in creating and maintaining energy and inspiration. Such leadership can come from both private and public actors, but our research showed that the characteristics necessary for getting a coalition together are possessed mainly by private 'leaders of change'. The personal characteristics of these leaders of change is one of the essential assets of a vital coalition. We discuss this and other conditions for vital coalitions in the next section.

7.3 Conditions for vital coalitions

The main thrust of our research was to identify conditions for vital coalitions at the regional level. In Chapter 3 we took the local urban conditions for vital coalitions as a starting point (Hendriks and Tops, 2002, 2005; Van de Wijdeven *et al.*, 2007).
- a sense of urgency;
- entrepreneurship;
- government backing.

Based on the findings in this book we suggest adding two further conditions: *a shared story line* and *versatile leadership*.

7.3.1 Sense of urgency and shared story line

In Chapter 3 we concluded that vital coalitions start out with a sense of urgency felt by the initiators of such coalitions. These initiators identify problems that, according to them, require immediate action. A sense of urgency was present in most cases, but this related mainly to the individual interests of the initiators of coalitions. For instance, in the Overdiepse polder case the individual farmers wanted to propose their own plan because they were afraid that otherwise their land would be expropriated in the near future. In the Sjalon case, falling prices for arable products, administrative burdens and limitations on farm development were the incentives for action by the farmers. Although the condition of a sense of urgency was met, the cases show that it was difficult to align this sense of urgency with the 'urgency' frames of other actors that were needed in the coalition to create a *shared sense of direction*. It was difficult to translate individual frames into a story line that would guide effective action.

In order to get things done, initiators will often try to elicit support from other actors, both public and private. The sense of urgency that was the reason for the initiative will have to be felt by these other actors too. As was illustrated in Chapter 6, issue framing and constructing an appealing story line is an important strategy for communicating a sense of urgency to potential supporters and build a vital coalition. Story lines were defined as mobilising forces that guide actions and bring people together in coalitions (Hajer, 1995; Rein and Schön, 1996). Story lines connect themes and issues to each other, and incorporate and adapt to changing events. They have the potential to reorder understandings and change policies (Rein and Schön, 1996). Initiators construct a story about what to achieve, why action is necessary and how the goals should be achieved so that they can connect to the frames of other actors they want to bring into or support the coalition. Story lines that connect ideas, people and resources are slowly built up during a process of interaction with these actors.

The New Markets case (Chapter 6) is a good illustration of the importance of a sense of urgency and constructing a shared story line. It was not possible to formulate an appealing and directional ambition for area development. On this regional scale it was difficult to align entrepreneurs in a business community which would invest in public goods, such as landscape quality. There was too much focus on coalition building and network management and too little on identifying issues and constructing problems, goals and strategies shared by the participants. Some entrepreneurs went their own way because they saw more opportunities to realise their individual ambitions in smaller and informal settings. There were no shared concerns or a sense of urgency to develop a collective vision on how new economic activities could contribute to landscape qualities. This meant there was no common ground for constructing an appealing story line to support such an ambition. The desire to design such a shared vision seemed to be held mainly by the experts involved and not by the entrepreneurs or government representatives.

In Chapter 6 Van der Stoep and Aarts pointed out that the construction of an appealing story line is probably the most important condition for vital coalitions. The composition of actors in coalitions, their resources and modes of alignment will always fluctuate. However, the agenda dimension of vital coalitions holds the potential for constructing a story line that is strong enough to carry and diffuse innovative ideas among potential supporters. They therefore suggest viewing the concept of vital coalitions more as a dynamic process of coalition building and storytelling than as a given group of people that can be characterised by more static descriptions in terms of the four dimensions of agenda, actors, modes of alignment and resources. Of course, these dimensions are highly relevant for identifying coalitions, but such descriptions will only be valid for a given moment in time and they do not explain change. If we want to learn about the conditions for successful vital coalitions and their contribution to sustainable regional development, we need to look at the processes or mechanisms behind these dimensions.

A condition related to such a process is *selective mobilisation* of actors. Selective mobilisation is the result of focusing on a sense of urgency and a particular story line that is appealing for some and not for others. The cases showed that coalitions will only be vital when participants really want to invest time, energy and money because they share some sense of urgency and can connect their individual goals to a collective story line. This is often not the case in interactive policy-making projects within traditional arrangements (see Chapter 2) where the main goal is to create consensus among stakeholders to prevent future protests. In such situations, innovation can be hampered because actors participate mainly to defend their interests (reactive), not because they are proactively inspired by a shared story line or a shared sense of urgency. That is why it is so difficult for process managers to establish a creative collective process that will yield innovative results. This is an important message in the Dutch context, in which governments often try to mobilise all relevant stakeholders in interactive regional processes, as we described in Chapter 2.

We can conclude from Chapter 4 that coalitions can be grouped into different categories of selective mobilisation. *Business-oriented coalitions* require entrepreneurial investors, *rural-urban coalitions* also contain citizens' representatives, while *steering coalitions* are characterised more by the participation of government authorities, often from different tiers of government. Expanding such coalitions with additional actors can have the advantage of additional support in the form of financial means, manpower, decision power, etc., but it may also imply a broadening of the agenda to satisfy the wishes of these new participants, and this might limit the productivity – or vitality – of the coalition.

7.3.2 Entrepreneurship and versatile leadership

Entrepreneurship is an important characteristic of pioneers, initiators or leaders of change in coalitions (see Chapter 5). In this respect, the role of entrepreneurship in regional areas is much the same as in urban areas, where private entrepreneurship has often been a key driving force for establishing vital coalitions (e.g. Hendriks and Tops, 2005; Van de Wijdeven *et al.*, 2007). Entrepreneurship exists in both the public and the private sector, but in the cases studied the innovative ideas and initiatives came from private entrepreneurs. Various innovation studies have also stressed that radical innovations usually come from outside the governing regime and are initially developed by entrepreneurs and pioneers (e.g. Constant, 1980; Utterback, 1994).

In Chapter 5, Horlings described several characteristics of leaders of change in the regional cases. She discussed how these leaders of change operated in the field of interaction between vital coalition and regime. In total, we can distinguish six characteristics of leaders of change:
1. availability of time;
2. open and flexible external attitude in networks and towards new knowledge;
3. good storytellers;
4. good networkers, in formal and informal contexts, functioning as boundary spanners in public-private partnerships;
5. acting in different roles and not afraid to act as a 'battering ram';
6. the ability to signal and use synchronicity of events.

First, *time* seems to be one of the most important resources of leaders of change. Most of them are in a phase of their lives where they are able to engage in a role as a pioneer of new initiatives. In our cases, time proved to be an important prerequisite for taking action, followed by inner motivation or more abstract incentives, such as awareness of environmental problems, corporate responsibility and appreciation of the value of the countryside.

Second, leaders of change have an *open and flexible external attitude*. Innovation occurs when entrepreneurs look for knowledge beyond the usual circles, for example by contacting and engaging with entrepreneurs in other sectors, and with knowledge institutes or universities.

Third, leaders of change create awareness and contribute to a shared sense of direction by *telling their story over and over again*. They try to convince others and 'vision between visions' in order to align people around a joint regional story line. They use a range of different competences, but a competence that seems to be somewhat underestimated in regional processes is the ability to generate a creative tension that makes people interested in development work and motivates them to take part, thus generating a sense of urgency (excitement capability).

Fourth, leaders of change are able to connect to formal and informal contexts. As networkers, leaders of change have to strike a balance between their own innovative network and the existing regime. They continually face the danger of incorporation into the institutional context or of losing touch with their own organisation or their business community. While good relations with various tiers of government are often needed to acquire the necessary administrative and/or political backing, leaders of change cannot distance themselves too far from the private actors in the coalition. Successful leaders of change have to know their way around in all these networks by using a variety of skills and competences, especially network power. They operate as 'boundary spanners' in public-private partnerships (see also Noble and Jones, 2006: 903). They contribute to bonding and bridging capital. The cases show that especially leaders who know the regional network very well are able to mobilise people and create commitment. Moreover, they show that leaders are connected and make use of both informal and formal networks of public and private actors. Informal co-operation offers opportunities for negotiations behind the scenes to create room for manoeuvre. Behind the scenes it is possible to speak freely, to influence people and to put new topics on the agenda.

Fifth, leaders of change know how to create strategic negotiation positions to achieve their goals. They work as visionaries and networkers, but also sometimes as a 'battering ram', threatening actions that would be unfavourable for the opponent (see Box 7.1 for an urban example). A much used strategy is trying to influence other public and private actors through the media, which can be used to overcome situations that seem to be in a deadlock or to acquire more supporters. Another strategy is using or threatening to use *less conventional means*. An example of this is pursuing a 'leapfrog strategy' to overcome a deadlock situation at the local level by trying to find policy support and common ground at a higher policy level, such as the provincial government or ministerial level.

Finally, the case of New Markets in Heuvelland in Chapter 6 pointed to the ability of leaders of change to *keep an eye open for synchronicity of events* which may lead to windows of opportunity for action. They are alert to new opportunities so that they can react quickly, but they also know how to be patient and wait for the right time to push their ideas. This ability to make use of or even 'arrange' coincidences is generally referred to as serendipity. So we may attribute a serendipitous characteristic to leaders of change.

Paradoxically, the success of leaders of change can also lead to their downfall. In general, dependence on one or a small number of individuals entails a risk. Sometimes, the coalition

> **Box 7.1. Entrepreneurship: acting as a 'battering ram'.**
>
> A confrontation between a civil servant and a leader of change (in the The Hague) who is not afraid to use less conventional means (Hendriks and Tops, 2005: 483).
>
> Civil servant: So, you lot are going to sweep.
> Leader: Yeah, isn't that marvellous.
> Civil servant: Yes, but you can't.
> Leader: Why not?
> Civil servant: You need a licence for that.
> Leader: And you are going to give me one.
> Civil servant: I don't think so.
> Leader: I think you will.
> Civil servant: If you start sweeping without a licence, you will be in breach, and I will be forced to call the police.
> Leader: By all means, go ahead. And when you do, I'll ask the press along. Just think of the headlines: police arrest citizens who want to clean their own street!
> Civil servant: And if you insist on sweeping the Weimarstraat, what are you going to do with the rubbish you've collected?
> Leader: We'll put it in the trade waste containers outside the shops.
> Civil servant: But that makes it commercial refuse, and you will have to pay a lot of money to get that dumped.
> Leader: Well, if that's so then perhaps we should simply dump this 'commercial refuse' in the garden of the alderman responsible for this mess. You know we've done it before, so if you don't start showing a teensy bit of cooperation, things may start to go very wrong indeed…

can be internally divided between those who support the line of the leader and those who have other opinions. Leaders of change may also move too quickly without ensuring that they have the full backing of their coalition or supporters, and become isolated. In the long run this may hamper the coalition's capacity to act. And if a leader of change for one reason or another gives up his or her function as the coalition front runner, the initiative may have a hard time continuing. Once the pioneering phase is over and the coalition enters a more stable phase, other competences are required of the leader and so it is important that a pioneering leader of change hands over his tasks to his successor in good time.

We conclude this section by noting that leadership has an inner and outer dimension, and an individual and collective dimension (see Wilber's psychological diagram, Table 5.3 in Chapter 5). In sustainable regional development studies the inner, individual dimension (the I dimension) in particular is still underestimated and tends to be neglected. We therefore suggest more research on the I dimension of sustainable regional development, focusing on the passion, values, motivations and driving forces of individual people.

7.3.3 Government backing

The last condition is that of government backing. Vital coalitions can only survive and fulfil (some of) their goals if governmental institutions or individuals in government with sufficient influence back the vital coalition or its leader of change. This does not always have to be a constant form of active support in the sense of high visibility or even the allocating of significant resources, but at the very least it has to be passive support for the coalition (a 'shadow of hierarchy' (Scharpf, 1997)) that provides the coalition with room for manoeuvre.

Government backing, or more particularly the lack of it, has led to some serious setbacks for the regional coalitions we investigated. Some of the most important hindering factors can be traced back to lack of government action (see Section 7.2 above). Most of these hindering factors are not attributable to one single governmental institution, let alone an individual; they are part of a more general encompassing structure which, in theory, can assist or facilitate vital coalitions, but which in reality often clashes with them.

Innovative concepts often contradict *current regulations*, as illustrated by the Northern Frisian Woods, Overdiepse polder and New Mixed Business cases. The area-based manure regulation proposed by the Northern Frisian Woods Association conflicted with national and European manure and fertiliser regulations. General procedures and rules, such as environmental impact assessments or the obligation to build manure storage facilities, often do not provide an adequate response to a transitional situation, as the case of the Overdiepse polder showed. The controversy around the New Mixed Business led to the imposition by government of additional sustainability criteria, to obtain the necessary permits.

Van der Stoep and Aarts note the *not invented here* syndrome with regard to the New Markets in Heuvelland case (Chapter 6). Several government agencies seemed reluctant to accept, or at least had their doubts about, external initiatives and potential vital coalitions formed elsewhere. The provincial government was much more amenable to initiatives that fitted into its own ideas, visions and plans. The authors explain that the provincial government was assigned a powerful role in regional development by the regional stakeholders and took this to be a confirmation of its own policy frames about regional development. Coalitions that were successful, such as the initiative for Regional Branding, were able to frame issues in terms of goals that were recognisable to the provincial government in the longer term. In this case, therefore, the province was perceived as an indispensable partner in the coalition. Because the provincial government was not challenged in its views about regional development, government backing could only be obtained by adapting initiatives to meet existing policy frames, which weakened the innovative potential of new ideas.

Furthermore, we saw a *collision of different regional agendas*, not only in Heuvelland, but also in the Sjalon case and Het Groene Woud. In the case of Het Groene Woud, ideas and initiatives fragmented. The branding initiative stimulated co-operation between actors and organisations, but in 2009 the tensions between nature conservation groups and entrepreneurs (the 'green' and 'red' tracks) continued and these groups did not become firmly allied around a joint agenda. In the Sjalon case we saw tensions between the agricultural agenda of the entrepreneurs and a more multifunctional-oriented governmental agenda.

Niche innovation often runs up against a *reluctant attitude* on the part of governments because the timing does not fit in with the planning process, the government authorities involved hold different opinions, the initiatives are considered too small, or the vital coalitions have not yet proven their sustainability credentials. Most initiatives have to make heavy investments of time, money and energy, not only in their own innovation but also in gaining political and social support.

Governmental backing is the most vulnerable aspect of a vital coalition, as it often proves hard to get government actors involved. In the next section, we analyse these problems further with help of the regime concept.

Comparing the cases described in this book against the conditions for vital coalitions discussed above, we conclude that two conditions seem to be lacking in most cases: a shared story line and government backing. While in most cases there is a clear sense of urgency and there are leaders of change eager to take the initiative, there is no common agenda to which actors are willing to subscribe. Moreover, several of the coalitions are not sufficiently supported by government actors or institutions.

7.4 The interrelations between the conditions

We have identified a number of conditions that have to be met for vital coalitions to be successful: a sense of urgency and a shared story line, entrepreneurship and versatile leadership, and government backing. These conditions do not stand on their own, but are closely linked. We use the following diagram (Figure 7.1) to illustrate the coherence and dynamics between these conditions. For instance, without a shared story line it is hard to get government backing, but a shared story line only emerges after a sense of urgency is established, and leaders of change play an important role in getting the backing for such a shared story line (see Figure 7.1).

The model clearly shows that building vital coalitions means carefully manoeuvring and balancing on the lines that connect a sense of urgency, a shared story line, entrepreneurship and leadership, and government backing. In Chapter 3 it was suggested that a balance is found

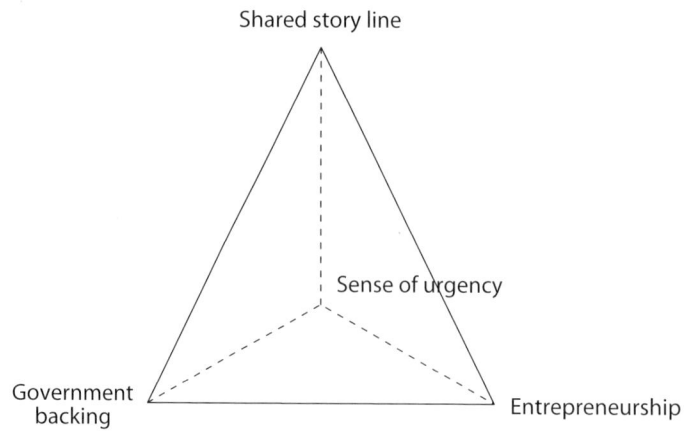

Figure 7.1. The interrelations between the conditions for vital coalitions.

when *institutional, situational, instrumental and cultural logics* are combined. Combining institutional and situational logics means combining the difference between applying universally applicable rules and procedures, which are often upheld by government actors and applied to concrete situations, and taking a concrete situation as the starting point for reasoning about what is necessary. In such situations, government backing means that governments are able to translate concrete ideas generated by vital coalitions into the institutional logic of permits, policy rules, land use plans, etc. Combining instrumental and cultural logics means not only having an eye for the rational problem-solving capacity of a vital coalition, but also knowing that the process in itself can be valuable and that this provides incentives for people to participate. The regional cases have relied on that mechanism and the regional vital coalitions have contained people who are able to combine the logic of governmental procedures with the action radius of citizens. The Overdiepse polder coalition, for instance, was able to shift between logics. It understood entrepreneurial problems, but because one of the members is also a member of an official committee overseeing the project, the coalition was also able to accommodate the logic of government procedures.

7.5 Regime-coalition interplay and reflection on the theory

7.5.1 No regional regime but 'regime dynamics'

Reflecting on regime-coalition interplay and the theory, it is by now apparent that we have applied the theory in a loose way. In our discussions on the use of the concept of urban regimes it became clear that a strict interpretation of the concept as a form of co-operation between business and government seems to relate more to North American urban practices (e.g. Van Ostaaijen, 2010). Nevertheless, the regional regime concept that we defined as 'the informal arrangements by which autonomous and semiautonomous actors work together to make and carry out governing decisions relevant to a specific region' has been useful in providing some analytical elements (agenda, coalition, resources, mode of alignment) to apply at the regional level.

An important insight is that regions, unlike cities, often do not have one encompassing agenda that guides actions and a clear coalition of public and private actors determining what is done. Regions deal with many authorities, coalitions and different agendas that serve diverse sectoral goals or diverging interests at various levels, from the national level to the local level. Compared to a municipality or a state, there is no clear public body connected to or responsible for the regional level (e.g. Van den Brink *et al.*, 2006). This has also been referred to as the 'administrative gap' that hampers the development of regions (see Chapter 2).

Although there may not be one regime at the regional level, there is often some form of institutionalised power at play, or some form of 'regime dynamics'. All the case studies show dominant policy practices and a semistructured set of rules and actors that influence the lower level. Sometimes there are attempts to organise co-operation between public organisations at the regional level. For example, in the New Markets in Heuvelland case (Chapter 6), urban and rural municipalities working together in the South Limburg Urban Network designed a collective agenda to guide that co-operation. Such co-operation between public actors aimed at collectively making governing decisions for the region can be considered a manifestation of regime dynamics.

The government actor that is most visible and influential in the region is the provincial government. Although provincial governments may pay only limited attention to regional issues because they are responsible for a much larger area, they do seem to be more suited than the smaller municipalities to taking on the role of front runner in regional development. Municipal authorities do not generally play this role because their prime concerns are their own municipal interests, a problem not uncommon in other regions and in other countries (e.g. Hamilton, 2002). On the regional level therefore, there is generally no common agenda and no strong coalition upholding and implementing this agenda; in other words, there is no (or no single) regional regime. Instead, there are multiple story lines that, at best, exist alongside each other, rather than reinforcing one another or coming together in one broadly supported regional agenda. Boxes 7.2 and 7.3 contain examples of the role of agendas in an urban context.

7.5.2 Relation between vital coalitions and regime dynamics

The urban examples show that the relation between a vital coalition and a regional regime can be diverse. The regime can hamper the development of the vital coalition, but also provide the necessary resources for it to function. In the Northern Frisian Woods, Sjalon, Het Groene Woud and Heuvelland cases, provincial, and in some cases municipal, governments in fact stimulated vital coalitions by providing subsidies or backing their agendas. Another situation arises when the goals of vital coalitions differ from those of the regime. Sometimes the main reason for their establishment is to resist current systemic structures that hamper alternative

---

Box 7.2. The benefits of a clear encompassing agenda: urban example I.

Between 2002 and 2006 Rotterdam's municipal government made safety its priority, with strong support from key actors in the city. Co-operation with the police and district attorney led to a strong coalition that could be called a regime (Tops, 2007; Van Ostaaijen, 2010). Most of the money available for 'new' policy between 2002 and 2006 was reserved for safety measures. The mayor was the political figurehead and a new local authority department co-ordinated the implementation of this priority in the policies for other local authority services and allocated funds. As a consequence, in the period between 2002 and 2006 there was a clear development towards safety policy as a comprehensive policy theme and discourse. All social, economic, planning and other policies had to take the safety perspective into account. For example, before 2002 the Housing Department, responsible for buying and selling the city's housing stock, had an economic perspective: making money. After 2002, the safety perspective was gradually introduced and the buying and selling of houses increasingly took account of the effects on neighbourhood safety (e.g. risk of deprivation). Even though this development towards safety as an overarching concern was never fully realised before political control of the city council changed in 2006, it was a clear policy focus between 2002 and 2006. The entire organisation knew that the municipal government's priority was safety and urban development policies were guided primarily by safety considerations. Government actors and citizens knew what to expect from the city council and private projects or citizens' initiatives that reflected this priority could count on considerable support (Van Ostaaijen, 2010).

> **Box 7.3. The absence of a clear encompassing agenda: urban example II.**
>
> After 2006, local government policies in Rotterdam were again pursued on more sectoral lines, a situation similar to that in Heuvelland. The new municipal government declared that both safety and a 'social programme' would be important. They decided to keep the 'safety pillar' (the discourse and organisational structure) intact, but added various social measures and instituted a 'social pillar': several social initiatives were clustered within a single overarching social programme, which was co-ordinated by a limited number of participants (executive councillors and local authority officers) and run in parallel to the 'safety pillar', which was the responsibility of the mayor. Both pillars developed their own logic and organisation and each had its own priorities. Co-ordination within the pillars was increasingly accompanied by co-ordination between the pillars, which led not only to more transaction (or 'complexity') costs, but also to confusion among the actors that do business with the city. An urban district officer noted that the policies of the social and safety pillar led to conflicting outcomes. For instance, the social pillar had made him remove a fence because it interfered with social objectives, while the safety pillar had led to the erection of the fence in the first place. Such confusion arises when there is no clear agenda to relate to at the regime level. In the Heuvelland case the division between an economic 'regime' (or discourse) and a more rural one may lead to similar problems. A single coherent agenda and coalition increases the focus and reduces complexity costs. It also makes it clearer to bottom-up initiatives in the region, such as vital coalitions, what the dominant discourse is, and therefore what conditions they have to meet to get support or what they need to resist.

development, for example if certain individuals or actors feel the current governing context (the regional regime) does not adequately respond to their local or regional needs.

Another possibility is that the absence of a capacity to act at the regime level leaves more room for local entrepreneurs and vital coalitions to step in and fill the void. Entrepreneurs, citizens and/or businesses can use this 'gap' to formulate their own agenda. In other words, a formal void can be offset by informal initiatives if vital coalitions have sufficient capacity to act to do that. However, the examples of coalitions in this research, although they sometimes successfully implemented innovative practices, did not challenge the dominant perspectives about regional development at the regime level.

To summarise, one of the main concepts for regional analysis in this research has been the regime concept. The original presumption that it would be possible to identify an encompassing agenda at the regional level was not tenable. In fact, we found that several agendas or story lines come into play at the regional level. However, this does not mean that there is no form of institutionalised power or regime dynamic functioning at the regional level. In all the cases, dominant policy practices, semistructured sets of rules, and informal and formal networks of actors together make up a regime dynamic, which facilitates as well as hinders the actions of potentially vital coalitions. In most cases, provincial governments can play a more facilitative role towards the vital coalitions by identifying trends and issues, taking their agendas on board, providing financial support, and especially by creating room for manoeuvre with regard to existing rules, procedures and regulations.

7.5.3 Regimes, niche innovation and agricultural transition

The results of this research can be compared with agricultural innovation research. Geels *et al.* (2008) carried out case studies on innovation in Dutch greenhouse horticulture and bioindustrial pork production based on the concept of strategic niche management. They describe how regime actors reproduce existing practices and operate within relatively stable rule-sets. There are existing tensions, but the regime remains stable as long as there is sufficient coherence between the regime actors (Grin and Van de Graaf, 1996). The authors describe transitions within Dutch agriculture as 'reconfiguration processes', which are more gradual processes than 'breakthrough transitions'. In these reconfiguration processes, which are driven by multiple innovation components, knowledge and innovation is transferred from developing actors to incumbent actors. Grin and Van de Graaf see niche innovation and experiments as very important mechanisms for changing the routines and working practices of farmers because in reconfiguration transitions 'old' actors are not replaced by new ones, but existing actors change their way of working by adopting niche innovation. Farmers or other parties may be active in niches and regimes at the same time, which corresponds with our findings.

Geels *et al.* (2008) see an important stimulating role in innovation for 'hybrid actors' who function between the 'insiders' and 'outsiders' of regimes. These hybrid actors stimulate the *anchoring* of the relation between niche and regime. Timing plays an important role in the process of anchoring. The study also suggests that more radical innovation can be stimulated in 'hybrid forums', where regime and niche developments come together at the most concrete level. The role of hybrid actors corresponds with the role of leaders of change in our research, who function as boundary spanners between their own network and the institutional context. Vital coalitions in which private and public actors work together can function as a forum where anchoring can take place between the niche innovation and the regime.

7.6 Recommendations

We make recommendations regarding vital coalitions as well as the regime level, encompassing bottom-up initiatives and the way they are hampered or stimulated by other actors at the regional level.

In the previous section we mentioned the administrative gap. An essential difference between the urban and regional level is that the latter in general lacks a strong public (and often also private) actor that possesses a legitimised governing capacity. At this level, the provincial government is suitably equipped to take on such a role as it is directly elected and has several powers regarding the development of the regions within its territory. Until now, though, the provincial authorities have not succeeded very well in bringing together different agendas to create shared, flexible and inspiring regional story lines that stimulate creativity and enable vital coalitions to develop and implement their ideas; to create a capacity to act.

The provincial government has a dual role. First, it can contribute to the construction of a shared story line as an anchor-point for different initiatives. For the province to take up such a co-ordinating role, a form of leadership at the governmental level is inevitable.

For regions characterised by multiple and continuously changing coalitions, an important leadership competence is network capability or 'network power', the ability to connect to these coalitions and link them together. This relies strongly on qualities such as personal interaction, communication and the ability to create trust. Another important condition for the provincial government to fulfil a more integrating role is to overcome the inner fragmentation of the provincial organisation, which is often divided into urban and rural policy areas.

Second, the province can take a more active role in identifying potential innovative initiatives and creating room for manoeuvre within the institutional context. In doing so, governments are in effect creating their own counterbalancing power. Even though the aims of new coalitions do not always coincide with governmental agendas, these coalitions are actual or potential forms of niche innovation, which may lead to new, more resilient regimes in the long run.

While a vital coalition is hard to realise from the top down, government authorities, particularly provincial governments, can adopt certain attitudes and internal procedures and policies that will enable them to create the conditions in which regional vital coalitions can flourish:

1. Civil servants from provincial and national government should be aware of innovative initiatives and vital coalitions emerging within the region, which implies that they must be given the time to do this. These civil servants can function as 'front liners' of the organisation and as boundary spanners between government and practice. They are not 'locked up' in public buildings, but have 'street knowledge' of emerging events and innovative coalitions (compare Lipsky's 'street level bureaucrats'(Lipsky, 1980), and for more on front line workers see Hartman and Tops 2005).
2. Government, civil servants and/or politicians can support these emerging vital coalitions by giving passive, verbal support or more active support, for example by providing money in the start-up phase, offering accommodation and taking a flexible approach to applying regulations. A supporting government is a proactive authority that plays a facilitative role, preferable organised outside the formal provincial organisation in a development organisations or programme office.
3. The provincial government should strike a balance between 'distance and commitment'. It should not suffocate or institutionalise initiatives, which stifles creativity and slows the process down. This often proves a difficult requirement for most government organisations as they often have a tendency to implement 'best practices' elsewhere, without realising that it is the specific local circumstances that make coalitions successful in the first place. The challenge facing government is not to interfere in the content of the coalition, but rather to provide a bedding within the institutional context without incorporating the initiative or smothering it with good intentions.

Vital coalitions have a loose bottom-up nature and are connected with acute local problems, and can be, and often are, strong centres of innovation. However, they face their own direction-setting problems, which limit their capacity to act. Coalitions are unavoidably embedded in institutional and planning contexts, and in order to realise their goals they often need some government support and have to relate to policy discourses. If they cannot find supporters for their goals within the policy context, they can (temporary) find support elsewhere, for example from scientific institutions or innovation programmes (such as TransForum or the InnovationNetwork in the Netherlands). It is also important for vital coalitions to recognise new windows of opportunity, for example when the key players change, new rules are introduced or policy agendas change.

Figure 7.1, which shows the interconnections between the different conditions for vital coalitions, can be a useful framework for determining where a vital coalition is being successful and what it still lacks. Vital coalitions emerge when all the conditions successfully work together. They also depend on inspiring leaders that are able to carry out an agenda and mobilise a coalition around it that is both inspiring and can contribute to regional development.

7.7 Summary

Finally, we briefly recap the conditions for vital coalitions that contribute to sustainable regional development in rural-urban regions.

We have identified three factors that contribute to the success of vital coalitions in the region:
- a sense of urgency and a shared story line;
- entrepreneurs and versatile leadership;
- government backing.

In many regions, the sense of urgency to make development more sustainable is clear to the stakeholders. However, regional sustainable story lines that describe the way forward and how to get there are still fragmented. We identified several coalitions at the regional level that are promising in terms of sustainable development, but public-private co-operation often proves to be difficult. The provinces are most suited to take on a more active leadership role and bring the different story lines together in a comprehensive agenda. A more active role in detecting, supporting and connecting vital coalitions is also desirable.

Vital coalitions can make valuable contributions to regional development through their innovative capacity. However, the paradox is that while success seems to depend, at least in part, on bottom-up initiatives (without top-down regulation), the absence of a regional story line or sense of direction leaves the region fragmented, and much energy lost in confusion or discussion about different interests with a variety of actors. Innovative entrepreneurs have to invest financially, socially and politically in new concepts, which can exceed their capabilities. In rural regions undergoing transition or where rural-urban boundaries are becoming blurred, establishing more sustainable regional agendas requires the innovative capacities and creativity of new vital coalitions.

References

Constant, E.W., 1980. The origins of the turbojet revolution. The Johns Hopkins University Press: Baltimore, MC, USA.

Geels, F., B. Elzen, E. Berkens, C. Leeuwis and B. Van Mierlo, 2008. Historische en hedendaagse systeeminnovaties in de glastuinbouw en varkenshouderij: Een innovatie-sociologische analyse. Eindrapport voor het Transforum project: Historical and future transitions in agriculture and food. TransForum, Zoetermeer, the Netherlands.

Grin, J. and H. Van de Graaf, 1996. Implementation as communicative action: An interpretive understanding of the interactions between policy makers and target groups. Policy Sciences 29(4), 291-319.

Hamilton, D., 2002. Regimes and regional governance: the case of Chicago. Journal of Urban Affairs 24(4): 403-423.

Hartman, C. and P.W. Tops, 2005. Frontlijnsturing: uitvoering op de publieke werkvloer van de stad. KCGS: the Hague, the Netherlands.

Hajer, M.A., 1995. The politics of environmental discourse: ecological modernization and the policy process. Clarendon Press: Oxford, UK.

Hendriks, F. and P.W. Tops, 2002. Het sloeg in als een bom: vitaal stadsbestuur en modern burgerschap in een Haagse stadsbuurt. Tilburg University: Tilburg, the Netherlands.

Hendriks, F. and P.W. Tops, 2005. Everyday Fixers as Local Heroes: A Case Study of Vital Interaction in Urban Governance. Local Government Studies 31(4): 475-490.

Kingdon, J.W., 2003. Agendas, alternatives, and public policies. Longman: New York, NY, USA.

Lipsky, M., 1980. Street-Level Bureaucracy. Dilemmas of the Individual in Public Services. Russell Sage Foundation, New York, NY, USA.

Noble, G. and R. Jones, 2006. The role of boundary-spanning managers in the Establishment of public-private partnerships. Public Administration 84(4): 891-917.

Utterback, J.M., 1994. Mastering the dynamics of innovation. Harvard Business School Press, Boston, MA, USA.

Rein, M., and D. Schön, 1996. Frame-Critical Policy Analysis and Frame-Reflective Policy Practice. Knowledge and Policy 9(1): 85-105.

Scharpf, F., 1997. Games Real Actors Play: Actor-Centered Institutionalism in Policy Research. Westview Press: Boulder, CO, USA.

Tops, P.W., 2007. Regimeverandering in Rotterdam: hoe een stadsbestuur zichzelf opnieuw uitvond. Atlas: Amsterdam, the Netherlands.

Van den Brink, A., A. Van der Valk and T. Van Dijk, 2006. Planning and the Challenges of the Metropolitan Landscape: Innovation in the Netherlands. International Planning Studies 11(3-4): 147-165.

Van der Valk, A. and T. Van Dijk (eds.), 2009. Regional Planning for Open Space. Routledge: London/New York, UK/USA.

Van de Wijdeven, T., E. Cornelissen, F. Hendriks and P.W. Tops, 2007. Vitale coalities: gewoon een kwestie van doen? over het ontstaan en voortbestaan van vitale coalities rond leefbaarheid in steden. In: E.M.H. Cornelissen P. Frissen and S. Kensen (eds.), Betoverend bestuur. Lemma: the Hague, the Netherlands, pp. 159-180.

Van Ostaaijen, J.J.C., 2010. Aversion and Accommodation. Political Change and Urban Regime Analysis in Dutch Local Government: Rotterdam 1998-2008. Dissertation research. Eburon: Delft, the Netherlands.

About the authors

Ina Horlings is currently working as a researcher in sustainable rural and regional development at the School of City and Regional Planning at Cardiff University (UK). She carried out this research while working at Telos, the Brabant Centre for Sustainable Development in Tilburg (the Netherlands). She previously worked at the Dutch Ministry of Agriculture, Nature and Food Quality on the national Agenda for the Living Countryside. She obtained her PhD in Policy Sciences and has published on several topics, including regional development, sustainability, self-organisation, policy, leadership and food production.
E-mail: horlingslg@cardiff.ac.uk

Hetty van der Stoep is a researcher and PhD candidate at the Land Use Planning Group of Wageningen University (the Netherlands). Her current research deals with the interaction between self-organising initiatives and governments and how this affects government agendas for the spatial development of metropolitan regions.
E-mail: hetty.vanderstoep@wur.nl

Julien van Ostaaijen is working as a researcher for the Tilburg School of Politics and Public Administration at Tilburg University (the Netherlands). His field of research is mainly local governance, urban developments in European and global cities, citizens participation, intra-municipal decentralisation and urban regimes. He has written a dissertation exploring the usefulness of urban regime analysis for Dutch local government and teaches an Urban Governance course.
E-mail: j.j.c.vanostaaijen@uvt.nl

Noelle Aarts is Professor of Strategic Communication at the University of Amsterdam and Associate Professor of Communication Strategies at Wageningen University (the Netherlands). Focusing on conversations between people, she studies and teaches inter-human processes and communication for creating space for change. She has published on several topics, including communication between organisations and their environment, negotiating environmental policies, competing claims in public space, dealing with ambivalence concerning farm animal welfare, and self-organisation and network building for regional innovation and multiple land use.
E-mail: noelle.aarts@wur.nl